STUDENT'S SOLUTIONS MANUAL
WILLIAM B. CRAINE III

to accompany

INTRO STATS

Richard D. De Veaux

Williams College

Paul F. Velleman

Cornell University

PEARSON
Addison
Wesley

Boston   San Francisco   New York
London   Toronto   Sydney   Tokyo   Singapore   Madrid
Mexico City   Munich   Paris   Cape Town   Hong Kong   Montreal

Reproduced by Pearson Addison-Wesley from electronic files supplied by the author.

Copyright © 2004 Pearson Education, Inc.
Publishing as Pearson Addison-Wesley, 75 Arlington Street, Boston MA 02116

ISBN   0-321-16590-x

2 3 4 5 6 CRS 06 05 04 03

Contents

Chapter 2 - Data

1. **The News.** Answers will vary.

3. **Oil spills.** *Who* – 50 recent oil spills. *What* – Spillage amount (no specified unit) and cause of puncture. *When* – Sometime between 1990 (Oil Pollution Act passed) and 1995 (report made in *Marine Technology*). *Where* – United States. *Why* – To determine whether or not spillage amount per oil spill has decreased since Congress passed the 1990 Oil Pollution Act and use that information in the design of new tankers. *How* – Not specified. *Cases* – Each of the 50 recent oil spills is a case. *Variables* – There are 2 variables. Spillage amount is a quantitative variable, and cause of puncture is a categorical variable.

5. **Weighing bears.** *Who* – 54 bears. *What* – Weight, neck size, length, and sex. *When* – Not specified. *Where* – Not specified. *Why* - Since bears are difficult to weigh, the researchers hope to use the relationships between weight, neck size, length, and sex of bears to estimate the weight of bears, given the other, more observable features of the bear. *How* – Researchers collected data on 54 bears they were able to catch. *Cases* – Each bear is a case.
Variables – There are 4 variables; weight, neck size, and length are quantitative variables, and sex is a categorical variable. No units are specified for the quantitative variables.
Concerns – The researchers are (obviously!) only able to collect data from bears they were able to catch. This method is a good one, as long as the researchers believe the bears caught are representative of all bears, in regard to the relationships between weight, neck size, length, and sex.

7. **Arby's menu.** *Who* – Arby's sandwiches. *What* – type of meat, number of calories, and serving size (in ounces). *When* – Not specified. *Where* – Arby's restaurants. *Why* – These data might be used to assess the nutritional value of the different sandwiches. *How* – Information was gathered from each of the sandwiches on the menu at Arby's, resulting in a census.
Cases – Each sandwich is a case. *Variables* – There are three variables. Number of calories and serving size are quantitative variables, and type of meat is a categorical variable.

9. **Babies.** *Who* – 882 births. *What* – Mother's age (in years), length of pregnancy (in weeks), type of birth (caesarean, induced, or natural), level of prenatal care (none, minimal, or adequate), birth weight of baby (unit of measurement not specified, but probably pounds and ounces), gender of baby (male or female), and baby's health problems (none, minor, major).
When – 1998-2000. *Where* – Large city hospital. *Why* – Researchers were investigating the impact of prenatal care on newborn health. *How* – It appears that they kept track of all births in the form of hospital records, although it is not specifically stated. *Cases* – Each of the 882 births is a case. *Variables* – There are three quantitative variables: mother's age, length of pregnancy, and birth weight of baby. There are four categorical variables: type of birth, level of prenatal care, gender of baby, and baby's health problems.

11. **Fitness.** *Who* – 25,892 men. *What* – Fitness level and cause of death. It is not clear what categories were used for fitness level. *When* - Over a 10-year period prior to the article being published in May 2002. *Where* – Not specified. *Why* – To establish an association between fitness level and death from cancer. *How* – Researchers tracked the group of men over a 10-year period. *Cases* – Each of the 25,892 men was an individual. *Variables* – There are two categorical variables: fitness level and cause of death.

13. Herbal medicine. *Who* – experiment volunteers. *What* – herbal cold remedy or sugar solution, and cold severity. *When* – Not specified. *Where* – Major pharmaceutical firm. *Why* – Scientists were testing the efficacy of an herbal compound on the severity of the common cold. *How* – The scientists set up a controlled experiment. *Cases* – Each volunteer is an individual. *Variables* – There are two variables. Type of treatment (herbal or sugar solution) is categorical, and severity rating is quantitative. *Concerns* – The severity of a cold seems subjective and difficult to quantify. Also, the scientists may feel pressure to report negative findings about the herbal product.

15. Cars. *Who* – Automobiles. *What* – Make, country of origin, type of vehicle and age of vehicle (probably in years). *When* – Not specified. *Where* – A large university. *Why* – Not specified. *How* – A survey was taken in campus parking lots. *Cases* – Each car is a case. *Variables* – There are three categorical variables and one quantitative variable. Make, country of origin, and type of vehicle are categorical variables, and age of vehicle is a quantitative variable.

17. Streams. *Who* – Streams. *What* – Name of stream, substrate of the stream (limestone, shale, or mixed), acidity of the water (measured in pH), temperature (in degrees Celsius), and BCI (unknown units). *When* – Not specified. *Where* – Upstate New York. *Why* – Research is conducted for an Ecology class. *How* – Not specified. *Cases* – Each stream is a case. *Variables* – There are five variables. Name and substrate of the stream are categorical variables, and acidity, temperature, and BCI are quantitative variables.

19. Air travel. *Who* – All airline flights in the United States. *What* – Type of aircraft, number of passengers, whether departures and arrivals were on schedule, and mechanical problems. *When* – This information is currently reported. *Where* – United States. *Why* – This information is required by the Federal Aviation Administration. *How* – Data is collected from airline flight information. *Cases* – Each flight is a case. *Variables* – There are four variables. Type of aircraft, departure and arrival timeliness, and mechanical problems are categorical variables, and number of passengers is a quantitative variable.

21. Refrigerators. *Who* – 41 refrigerators. *What* – Brand, cost (probably in dollars), size (in cu. ft.), type, estimated annual energy cost (probably in dollars), overall rating, and repair history (in percent requiring repair over the past five years). *When* – 2002. *Where* – United States. *Why* – The information was compiled to provide information to the readers of *Consumer Reports*. *How* – Not specified. *Cases* – Each of the 41 refrigerators is a case. *Variables* – There are 7 variables. Brand, type, and overall rating are categorical variables. Cost, size, estimated energy cost, and repair history are quantitative variables.

Chapter 3 – Displaying Categorical Data

1. **Graphs in the news.** Answers will vary.

3. **Tables in the news.** Answers will vary.

5. **Causes of death.**

   a) Yes, it is reasonable to assume that heart and respiratory disease caused approximately 38% of U.S. deaths in 1999, since there is no possibility for overlap. Each person could only have one cause of death.

   b) Since the percentages listed add up to 73.7%, other causes must account for 26.3% of US deaths.

   c) A pie chart is a good choice (with the inclusion of the "Other" category), since causes of US deaths represent parts of a whole. A bar chart would also be a good display.

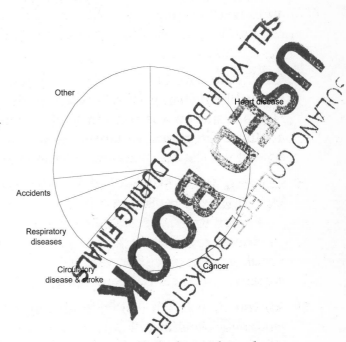

7. **Ghosts.**

   a) It is NOT reasonable to assume the 66% of those polled expressed a belief in either ghosts or astrology. The percentages in the table add up to 185%! This tells us that we are not dealing with parts of a whole and that some respondents believe in more than one of the psychic phenomena listed. In other words, belief in ghosts and belief in astrology are not mutually exclusive. There is no way to know what percent of respondents believe in ghosts or astrology.

   b) As in 7a, since the percentages are not meant to add up to 100%, there is no way to determine what percentage of respondents did not believe in any of these phenomena. With data collected in this fashion, the only way to determine the percentage of people who did not believe in any of the psychic phenomena would be to add another category to the survey when collecting the data.

   c) Since the percentages were not intended to add up to 100%, a pie chart is not appropriate. A bar chart nicely displays the percentages as relative heights of bars. Also, we are probably interested in the most popular phenomena, it makes sense to make a Pareto Chart, with the most common belief occurring first in the chart.

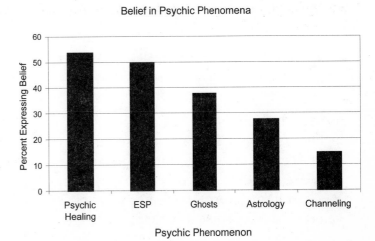

9. Oil spills.

The bar chart shows that grounding is the most frequent cause of oil spillage for these 50 spills, and allows the reader to rank the other types as well. If being able to differentiate between these close counts is required, use the bar chart. The pie chart is also acceptable as a display, but it's difficult to tell whether, for example, there is a greater percentage of spills caused by grounding or hull failure. If you want to showcase the causes of oil spills as a fraction of all 50 spills, use the pie chart.

11. Teens and technology.

a) 67% of teens use a calculator and 46% of teens use an answering machine, for a difference of 21%

b) The display makes it appear that more than twice as many teens use calculators as use answering machines.

c) The vertical scale should display percentages starting at 0. This isn't simply an improvement, but rather a necessity. The display provided is misleading.

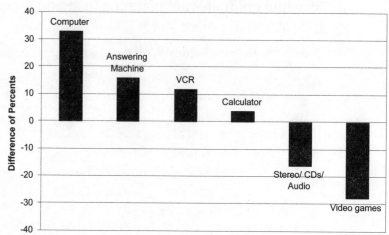

d) The display at the right is the Importance Gap in teens' opinions about technology. It displays the differences in percentages of teens who stated that the piece of technology was critical to use and the percentage of teens who actually own the piece of technology.

e) Teens tend to think that some pieces of technology are critically important to own although they don't use them daily. The Importance Gap display highlights the fact that computers and video games fall on the two extremes of this difference. 77% of teens feel computers are important to own, yet fewer than half use them daily.

13. Auditing reform.

a) The pie chart at the right displays the same information as the bar chart.

b) The bar chart makes it easier to compare the percentages. It also preserves the order in which the questions were originally asked.

c) Most respondents in a Gallup poll conducted February 8-10, 2002 favored some sort of reform in the way that corporations are audited. 39% of the respondents advocated major reforms, 35% said minor reforms were needed, and 17% favored a complete overhaul of the corporate auditing system. Only 4% were in favor of no reforms to the current system.

What Kind of Changes Are Needed in the Way Corporations Are Audited?

15. Class of 2000.

a) A table with marginal totals is to the right. There are 268 White graduates and 325 total graduates. 268/325 ≈ 82.5% of the graduates are White.

Plans	White	Minority	TOTAL
4-year college	198	44	242
2-year college	36	6	42
Military	4	1	5
Employment	14	3	17
Other	16	3	19
TOTAL	**268**	**57**	**325**

b) There are 42 graduates planning to attend 2-year colleges. 42/325 ≈ 12.9%

c) 36 white graduates are planning to attend 2-year colleges. 36/325 ≈ 11.1%

d) 36 white graduates are planning to attend 2-year colleges and there are 268 whites graduates. 36/268 ≈ 13.4%

e) There are 42 graduates planning to attend 2-year colleges. 36/42 ≈ 85.7%

f) A segmented bar chart is a good display of these data:

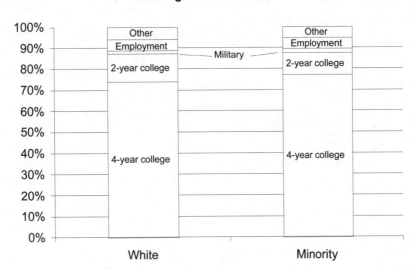

g) The conditional distributions of plans for Whites and Minorities are similar: White – 74% 4-year college, 13% 2-year college, 2% military, 5% employment, 6% other. Minority – 77% 4-year college, 11% 2-year college, 2% military, 5% employment, 5% other. Caution should be used with the percentages for Minority graduates, because the total is so small. Each graduate is almost 2%. Still, the conditional distributions of plans are essentially the same for the two groups. There is little evidence of an association between race and plans for after graduation.

17. Canadian languages.

a) 19,134,000 Canadians speak English only. 19,134,000/28,529,000 total Canadians ≈ 67.1%

b) 4,078,000 Canadians speak French only and 4,843,000 speak both French and English, for a total of 8,921,000 French speakers. 8,921,000/28,529,000 total Canadians ≈ 31.3%

c) 3,952,000 French and 2,661,000 French and English speakers yield a total of 6,613,000 French speakers in Quebec. 6,613,000/7,045,000 Quebec residents ≈ 93.9%

d) 6,613,000 Quebec residents speak French and 8,921,000 Canadians speak French. The percentage of French-speaking Canadians who live in Quebec is 6,613,000/8,921,000 ≈ 74.1%

e) If language knowledge were independent of Province, we would expect the percentage of French-speaking residents of Quebec to be the same as the overall percentage of Canadians who speak French. Since 31.3% of all Canadians speak French while 93.9% of residents of Quebec speak French, there is evidence of an association between language knowledge and Province.

19. Weather forecasts.

a) The table shows the marginal totals. It rained on 34 of 365 days, or 9.3% of the days.

b) Rain was predicted on 90 of 365 days. 90/365 ≈ 24.7% of the days.

		Actual Weather		Total
		Rain	No Rain	
Forecast	Rain	27	63	90
	No Rain	7	268	275
	Total	34	331	365

c) The forecast of rain was correct on 27 of the days it actually rained and the forecast of No Rain was correct on 268 of the days it didn't rain. So, the forecast was correct a total of 295 times. 295/365 ≈ 80.8% of the days.

d) There is evidence of an association between the type of weather and the ability of the forecasters to make an accurate prediction. When Rain was forecast, it actually rained only 27 out of 90 times (30%). When the forecasters predicted No Rain, they were correct an amazing 268 out of 275 times (97.5%). The weather forecasters' predictions of Rain are much less reliable than their predictions of No Rain. However, this is hardly a ringing endorsement of the weather forecasters abilities to predict No Rain. Since the

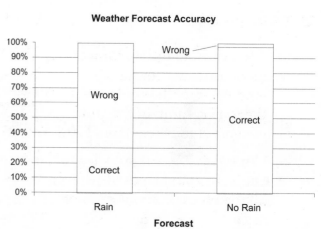

actual weather is No Rain on 90.7% of the days, these predictions are a pretty safe bet.

21. Working parents.

Which do you see as the ideal family?

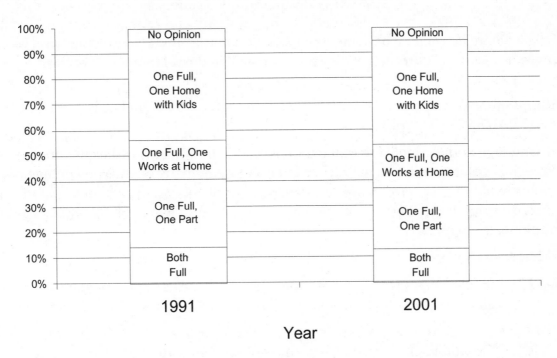

The Gallup poll doesn't provide strong evidence of a change in people's opinions regarding the ideal family in today's society between the years of 1991 and 2001. The conditional distributions of opinion by year appear roughly the same. For example, the percentage of respondents in 1991 who thought the ideal family had two parents that worked fulltime was 14%, and in 2001, the percentage was 13%.

23. Blood pressure.

a) The marginal distribution of blood pressure for the employees of the company is the total column of the table, converted to percentages. 20% low, 49% normal and 31% high blood pressure.

Blood pressure	under 30	30 - 49	over 50	Total
low	27	37	31	95
normal	48	91	93	232
high	23	51	73	147
Total	98	179	197	474

b) The conditional distribution of blood pressure within each age category is:
Under 30 : 28% low, 49% normal, 23% high
30 – 49 : 21% low, 51% normal, 28% high
Over 50 : 16% low, 47% normal, 37% high

c) A segmented bar chart of the conditional distributions of blood pressure by age category is at the right.

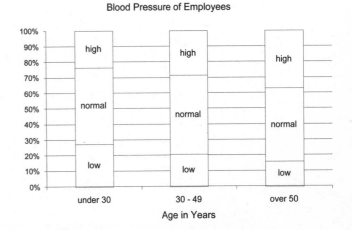

Blood Pressure of Employees

d) In this company, as age increases, the percentage of employees with low blood pressure decreases, and the percentage of employees with high blood pressure increases.

e) No, this does not prove that people's blood pressure increases as they age. Generally, an association between two variables does not imply a cause-and-effect relationship. Specifically, these data come from only one company and cannot be applied to all people. Furthermore, there may be some other variable that is linked to both age and blood pressure. Only a controlled experiment can isolate the relationship between age and blood pressure.

25. Family planning.

The percentage of unplanned pregnancies decreased as education level increased. However, this does *not* mean that more schooling taught young women better family planning. Association between variables is not evidence of a cause-and-effect relationship. In this case, other socioeconomic variables might be related to both unplanned pregnancy and education level. Perhaps some women even had to leave school *because* of an unplanned pregnancy.

27. Worldwide toy sales.

a) 2042.16 million dollars worth of toys are sold in European department stores, and 5987.6 million dollars worth of toys are sold in all department stores. 2042.16/5987.6 ≈ 34.1%

b) 2843.8 million dollars worth of toys are sold through catalogs. The "World" total of the table is 74395 million dollars worth of toys sold. The percentage of all toys sold through catalogs is 2843.8/74395 ≈ 3.8%.

c) Toy Chains are more popular sellers of toys in Europe than in North America, as are Toy, Hobby and Game retailers and Department Stores. In North America, General Merchandise and Food, Drug and Misc. Outlets are more popular than they are in Europe.

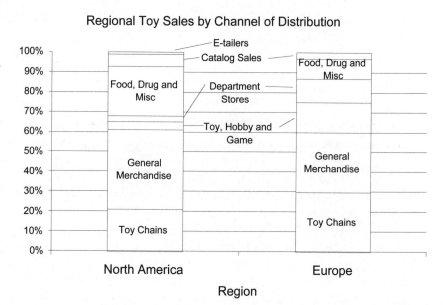

d) The distribution channel varies by region, although General Merchandise stores are either the most popular or a close second in all regions. Toy Chains are especially important in Europe and Toy, Hobby and Game Shops are important in Asia. Only North America has a significant percentage of toys sold in Food and Drug Stores. In the year 1999, E-tailing of toys was of little significance.

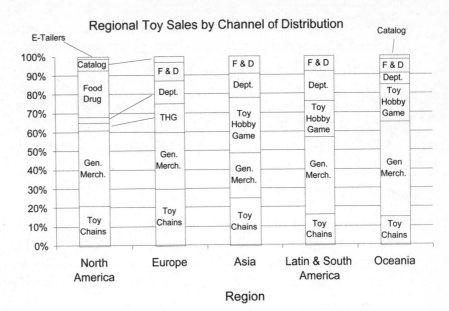

29. Hospitals.

a) The marginal totals have been added to the table:

		Discharge delayed		
		Large Hospital	**Small Hospital**	**Total**
Procedure	**Major surgery**	120 of 800	10 of 50	130 of 850
	Minor surgery	10 of 200	20 of 250	30 of 450
	Total	130 of 1000	30 of 300	160 of 1300

160 of 1300, or about 12.3% of the patients had a delayed discharge.

b) Major surgery patients were delayed 130 of 850 times, or about 15.3% of the time. Minor Surgery patients were delayed 30 of 450 times, or about 6.7% of the time.

c) Large Hospital had a delay rate of 130 of 1000, or 13%.
Small Hospital had a delay rate of 30 of 300, or 10%.
The small hospital has the lower overall rate of delayed discharge.

d) Large Hospital: Major Surgery 15% delayed and Minor Surgery 5% delayed.
Small Hospital: Major Surgery 20% delayed and Minor Surgery 8% delayed.
Even though small hospital had the lower overall rate of delayed discharge, the large hospital had a lower rate of delayed discharge for each type of surgery.

e) No. While the overall rate of delayed discharge is lower for the small hospital, the large hospital did better with *both* major surgery and minor surgery.

f) The small hospital performs a higher percentage of minor surgeries than major surgeries. 250 of 300 surgeries at the small hospital were minor (83%). Only 200 of the large hospital's 1000 surgeries were minor (20%). Minor surgery had a lower delay rate than major surgery (6.7% to 15.3%), so the small hospital's overall rate was artificially inflated. Simply put, it is a mistake to look at the overall percentages. The real truth is found by looking at the rates after the information is broken down by type of surgery, since the delay rates for each type of surgery are so different. The larger hospital is the better hospital when comparing discharge delay rates.

31. Graduate admissions.

a) 1284 applicants were admitted out of a total of 3014 applicants. 1284/3014 = 42.6%

b) 1022 of 2165 (47.2%) of males were admitted. 262 of 849 (30.9%) of females were admitted.

Program	Males Accepted (of applicants)	Females Accepted (of applicants)	Total
1	511 of 825	89 of 108	600 of 933
2	352 of 560	17 of 25	369 of 585
3	137 of 407	132 of 375	269 of 782
4	22 of 373	24 of 341	46 of 714
Total	1022 of 2165	262 of 849	1284 of 3014

c) Since there are four comparisons to make, the table at the right organizes the percentages of males and females accepted in each program. Females are accepted at a higher rate in every program.

Program	Males	Females
1	61.9%	82.4%
2	62.9%	68.0%
3	33.7%	35.2%
4	5.9%	7%

d) The comparison of acceptance rate within each program is most valid. The overall percentage is an unfair average. It fails to take the different numbers of applicants and different acceptance rates of each program. Women tended to apply to the programs in which gaining acceptance was difficult for everyone. This is an example of Simpson's Paradox.

Chapter 4 – Displaying Quantitative Data

1. **Statistics in print.** Answers will vary.

3. **Thinking about shape.**

    a) The distribution of the number of speeding tickets each student in the senior class of a college has ever had is likely to be unimodal and skewed to the right. Most students will have very few speeding tickets (maybe 0 or 1), but a small percentage of students will likely have comparatively many (3 or more?) tickets.

    b) The distribution of player's scores at the U.S. Open Golf Tournament would most likely be unimodal and slightly skewed to the right. The best golf players in the game will likely have around the same average score, but it is easier to score 15 strokes below the mean than 15 strokes above it.

    c) The weights of female babies in a particular hospital over the course of a year will likely have a distribution that is unimodal and symmetric. Most newborns have about the same weight, with some babies weighing more and less than this average. There may be slight skew to the left, since there seems to be a greater likelihood of premature birth (and low birth weight) than post-term birth (and high birth weight).

    d) The distribution of the length of the average hair on the heads of students in a large class would likely be bimodal and skewed to the right. The average hair length of the males would be at one mode, and the average hair length of the females would be at the other mode, since women typically have longer hair than men. The distribution would be skewed to the right, since it is not possible to have hair length less than zero, but it is possible to have a variety of lengths of longer hair.

5. **Heart attack stays.**

    The distribution of the length of hospital stays of female heart attack patients is skewed to the right, with stays ranging from 1 day to 34 days. The distribution is centered around 8 days, with the majority of the hospital stays lasting between 1 and 15 days. There are a relatively few hospital stays longer than 27 days. Many patients have a stay of only one day, possibly because the patient died.

7. **Sugar in cereals.**

    a) The distribution of the sugar content of breakfast cereals is bimodal, with a cluster of cereals with sugar content around 10% sugar and another cluster of cereals around 48% sugar. The lower cluster shows a bit of skew to the right. Most cereals in the lower cluster have between 0% and 10% sugar. The upper cluster is symmetric, with center around 45% sugar.

    b) There are two different types of breakfast cereals, those for children and those for adults. The children's cereals are likely to have higher sugar contents, to make them taste better (to kids, anyway!). Adult cereals often advertise low sugar content.

9. Wineries.

a) There is information displayed about 36 wineries and it appears that 25 of the wineries are smaller than 50 acres. That's around 69% of the wineries. (75% would be a good estimate!)

b) The distribution of the size of 36 Finger Lakes wineries is skewed to the right. Most wineries are smaller than 75 acres, with a few larger ones, from 75 to 175 acres. One winery was larger than all the rest, over 250 acres. The mode of the distribution is between 25 and 50 acres.

11. Home runs.

The distribution of the number of homeruns hit by Mark McGwire during the 1986 – 2000 seasons is skewed to the right, with a typical number of homeruns per season in the 30s. With the exception of 3 seasons in which McGwire hit fewer than 10 homeruns, his total number of homeruns per season was between 22 and the maximum of 70.

13. Horsepower.

The distribution of horsepower of cars reviewed by _Consumer Reports_ is nearly uniform. The lowest horsepower was 65 and the highest was 155. The center of the distribution was around 105 horsepower.

```
Consumer
 Reports
Horsepower

15 | 05
14 | 2
13 | 0358
12 | 0559
11 | 00555
10 | 359
 9 | 00577       KEY:
 8 | 0058       11 | 5 =
 7 | 01158     115 horsepower
 6 | 55889
```

15. Hurricanes.

a) A dotplot of the number of hurricanes each year from 1944 through 2000 is displayed at the right. Each dot represents a year in which there were that many hurricanes.

b) The distribution of the number of hurricanes per year is unimodal and skewed to the right, with center around 2 hurricanes per year. The number of hurricanes per year ranges from 0 to 7. There are no outliers. There may be a second mode at 5 hurricanes per year, but since there were only 4 years in which 5 hurricanes occurred, it is unlikely that this is anything other than natural variability.

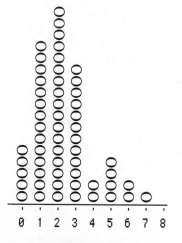

Number of Hurricanes

17. Acid rain.

The distribution of the pH readings of water samples in Allegheny County, Penn. is bimodal. A roughly uniform cluster is centered around a pH of 4.4. This cluster ranges from pH of 4.1 to 4.9. Another smaller, tightly packed cluster is centered around a pH of 5.6. Two readings in the middle seem to belong to neither cluster.

Acidity of Water Samples in Allegheny County, Penn.

19. Hospital stays.

a) The histograms of male and female hospital stay durations would be easier to compare if the were constructed with the same scale, perhaps from 0 to 20 days

b) The distribution of hospital stays for men is skewed to the right, with many men having very short stays of about 1 or 2 days. The distribution tapers off to a maximum stay of approximately 25 days. The distribution of hospital stays for women is skewed to the right, with a mode at approximately 5 days, and tapering off to a maximum stay of approximately 22 days. Typically, hospital stays for women are longer than those for men.

c) The peak in the distribution of women's hospital stays can be explained by childbirth. This time in the hospital increases the length of a typical stay for women, and not for men.

21. Final grades.

The width of the bars is much too wide to be of much use. The distribution of grades is skewed to the left, but not much more information can be gathered.

23. Zipcodes.

Even though zipcodes are numbers, they are not quantitative in nature. Zipcodes are categories. A histogram is not an appropriate display for categorical data. The histogram the Holes R Us staff member displayed doesn't take into account that some 5-digit numbers do not correspond to zipcodes or that zipcodes falling into the same classes may not even represent similar cities or towns. The employee could design a better display by constructing a bar chart that groups together zipcodes representing areas with similar demographics and geographic locations.

25. Productivity study.

The productivity graph is useless without a horizontal scale indicating the time period over which the productivity increased. Also, we don't know the units in which productivity is measured.

27. Law enforcement.

a) The histograms at the right show the number of assaults per 1000 officers and number of killed or injured per 1000 officers, for eleven federal agencies with officers authorized to carry firearms and make arrests.

Assaults
(per 1000)

Killed – Injured
(per 1000)

b) The distribution of assault rates for these federal agencies is roughly symmetric with two high outliers. The center of the distribution is between 5 and 10 assaults per 1000 officers and, with the exception of two agencies, BATF and National Park Service, with over 30 assaults per 100 officers, the distribution is tightly clustered. The distribution of killed and injured rates for these eleven law enforcement agencies is very tightly clustered with almost all of the agencies reporting rates of between 0 and 5 officers per 1000 killed and injured. Customs Service, with a rate of 5.1 officers per 1000, is essentially part of this group, as well. There is one high outlier, the National Park Service, with a rate of 15 officers per 1000 killed and injured.

c) The National Park Service is a high outlier for assaults and killed-injured, with rates of 38.7 officers per 1000 assaulted and 15 officers per 1000 killed or injured. The BATF is a high outlier for assaults, with 31.1 officers per 1000 assaulted.

29. MPG.

a) A back-to-back stemplot of these data is shown at the right. A plot with with tens and units digits for stems and tenths for leaves would have been quite long, but still useable. A plot with tens as stems and rounded units as leaves would have been too compact. This plot has tens as the stems, but the stems are split 5 ways. The uppermost 2 stem displays 29 and 28, the next 2 stem displays 27 and 26, and so on. The key indicates the rounding used, as well as the accuracy of the original data. In this case, the mileages were given to the nearest tenth.

US Cars		Other Cars
	3	7
44	3	45
	3	222
01	3	01
89	2	8
777	2	7
	2	
2	2	222
11	2	0
888999	1	
6777	1	67

KEY:
2 | 5 =
24.5 – 25.4 mpg

b) In general, the Other cars got better gas mileage than the US Cars, although both distributions were highly variable. The distribution of US cars was bimodal and skewed to the right, with many cars getting mileages in the high teens and low twenties, and another group of cars whose mileages were in the high twenties and low thirties. Two high outliers had mileages of 34 miles per gallon. The distribution of Other cars, in contrast, was bimodal and skewed to the left. Most cars had mileages in the high twenties and thirties, with a small group of cars whose mileages were in the low twenties. Two low outliers had mileages of 16 and 17 miles per gallon.

31. Nuclear power.

a) The stemplot at the right shows the distribution of construction costs, measured in $1000 per mW of 12 nuclear power generators.

b) The distribution of nuclear power plant construction costs is skewed left, and has several modes, with gaps in between. Several plants had costs near $80,000 per mW, another few had costs near $60,000 per mW and several had costs near $30,000 per mW.

```
Construction Cost
   ($1000/mW)

8 | 0148
7 | 9
6 | 023      Key:
5 | 6         7|9 = $79000
4 |              per mW
3 | 25
2 | 8
```

c) The timeplot at the right shows the change in nuclear plant construction costs over time.

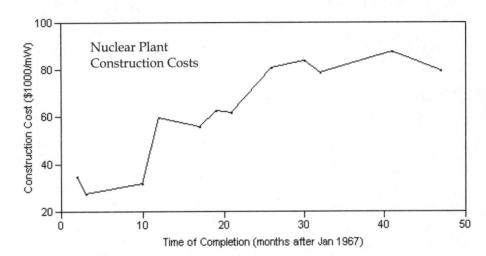

d) From the timeplot, it is apparent that the nuclear plant construction costs generally increased over time.

33. Assets.

a) The distribution of assets of 79 companies chosen from the *Forbes* list of the nation's top corporations is skewed so heavily to the right that the vast majority of companies have assets represented in the first bar of the histogram, 0 to 10 billion dollars. This makes meaningful discussion of center and spread impossible.

b) The distribution of logarithm of assets is preferable, because it is roughly unimodal and symmetric. The distribution of the square root of assets is still skewed right, with outliers.

c) If $\sqrt{Assets} = 50$, then the companies assets are approximately $50^2 = 2500$ million dollars.

d) If $\log(Assets) = 3$, then the companies assets are approximately $10^3 = 1000$ million dollars.

Chapter 5 – Describing Distributions Numerically

1. **In the news.** Answers will vary.

3. **Summaries.**

   a) The mean price of the electric smoothtop ranges is $1001.50.

   b) In order to find the median and the quartiles, the list must be ordered.
   565   750   850   900   1000 1050 1050 1200 1250 1400
   The median price of the electric ranges is $1025.
   Quartile 1 = $850 and Quartile 3 = $1200.

   c) The range of the distribution of prices is Max – Min = $1400 – $565 = $835.
   The IQR = Q3 – Q1 = $1200 - $850 = $350.

5. **Mistake.**

   a) As long as the boss's true salary of $200,000 is still above the median, the median will be correct. The mean will be too large, since the total of all the salaries will decrease by $2,000,000 - $200,000 = $1,800,000, once the mistake is corrected.

   b) The range will likely be too large. The boss's salary is probably the maximum, and a lower maximum would lead to a smaller range. The IQR will likely be unaffected, since the new maximum has no effect on the quartiles. The standard deviation will be too large, because the $2,000,000 salary will have a large squared deviation from the mean.

7. **Payroll.**

   a) The mean salary is $\dfrac{(1200 + 700 + 6(400) + 4(500))}{12} = \$525$.

   The median salary is the middle of the ordered list:
   400   400   400   400   400   400   500   500   500   500   700   1200
   The median is $450.

   b) Only two employees, the supervisor and the inventory manager, earn more than the mean wage.

   c) The median better describes the wage of the typical worker. The mean is affected by the two higher salaries.

   d) The IQR is the better measure of spread for the payroll distribution. The standard deviation and the range are both affected by the two higher salaries.

9. **Standard deviation.**

   a) Set 2 has the greater standard deviation. Both sets have the same mean, 6, but set two has values that are generally farther away from the mean.
   SD(Set 1) = 2.24   SD(Set 2) = 3.16

   b) Set 2 has the greater standard deviation. Both sets have the same mean (15), maximum (20), and minimum (10), but 11 and 19 are farther from the mean than 14 and 16.
   SD(Set 1) = 3.61   SD(Set 2) = 4.53

c) The standard deviations are the same. Set 2 is simply Set 1 + 80. Although the measures of center and position change, the spread is exactly the same.
SD(Set 1) = 4.24   SD(Set 2) = 4.24

11. Home runs.

5-number summary: 8, 14, 24.5, 33, 61   IQR = Q3 – Q1 = 33 – 14 = 19 home runs.

Using the Outlier Rule (more than 1.5 IQRs beyond the quartiles) to find the fences:

Upper Fence: $Q3 + 1.5(IQR) = 33 + 1.5(19)$
$$= 33 + 28.5$$
$$= 61.5$$

Lower Fence: $Q1 - 1.5(IQR) = 14 - 1.5(19)$
$$= 14 - 28.5$$
$$= -14.5$$

Home Runs

Any year with more than 61.5 home runs or less than –14.5 home runs (a meaningless fence) is considered an outlier. Technically, the year in which Maris hit 61 home runs is **not** an outlier, but according to a less formal definition (any point that departs from the overall pattern), it seems strange. For this reason, 61 homeruns is displayed as an outlier in the boxplot.

13. Marriage age.

The distribution of marriage age of U.S. men is skewed right, with a typical man (as measured by the median) first marrying at around 24 years old. The middle 50% of male marriage ages is between about 23 and 26 years. For U.S. women, the distribution of marriage age is also skewed right, with median of around 21 years. The middle 50% of female marriage age is between about 20 and 23 years. When comparing the two distributions, the most striking feature is that the distributions are nearly identical in spread, but have different centers. Females typically seem to marry earlier than males. In fact, between 50% and 75% of the women marry at a younger age than *any* man.

15. Wines.

a) A Seneca Lake winery had the maximum case price of approximately $150.

b) A Seneca Lake winery had the minimum case price of approximately $50.

c) The Keuka Lake wineries were consistently high.

d) Cayuga Lake wineries and Seneca Lake wineries have approximately the same average case price, at about $100, while a typical Keuka Lake winery has a case price of about $130. Keuka Lake wineries have consistently high case prices, between $120 and $140, with one low outlier at about $85 per case. Cayuga Lake wineries have case prices from $70 to $135, and Seneca Lake wineries have highly variable case prices, from $50 to $150.

17. Unemployment.

The distribution of unemployment rates reported by the U.S. Bureau of Labor Statsitics must have had about 30% of the rates above 4.2% and the rest below. In other words, 4.2% should be slightly below the third quartile (25% of rates above). Other than that, the distribution cannot be accurately determined. An example of one possible distribution is shown in the boxplot at the right.

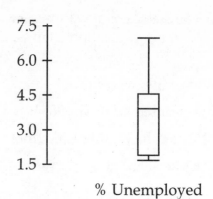

% Unemployed

19. Test scores.

a) Class 3 had the highest mean score, probably somewhere in the 70s. The other two classes had mean scores in the 60s.

b) Class 3 had the highest median score, probably somewhere in the 80s. The other two classes had median scores in the 60s.

c) Class 3 has median higher than the mean, since the distribution of scores is skewed to the left. The mean is pulled toward the tail. The other two classes have roughly symmetric distributions, with mean and median approximately equal.

d) Class 1 has the smallest standard deviation. Most scores are clustered close to the mean.

e) Class 1 probably has the smallest IQR. However, without the actual scores, it is impossible to calculate the exact IQR of these classes. We can estimate them. Class 1 has 24 students, so Q1 is between the 6th and 7th scores, somewhere between 50 and 60. Q3 is between the 18th and 19th scores, somewhere between 70 and 80. The IQR is at least 10 and at most 30. Class 3 also has 24 scores. Q1 is between 60 and 70. Q3 is between the 18th score (80-90) and the 19th score (90-100), meaning it could be between 80 and 100. The IQR is at least 10 and at most 40. Class 2 seems to have the largest IQR, of at least 30 and at most 50.

21. Still rockin'.

a) The histogram and boxplot of the distribution of "crowd crush" victims' ages both show that a typical crowd crush victim was approximately 18 - 20 years of age, that the range of ages is 36 years, that there are two outliers, one victim at age 36 - 38 and another victim at age 46 – 48.

b) This histogram shows that there may have been two modes in the distribution of ages of "crowd crush" victims, one at 18 - 20 years of age and another at 22 – 24 years of age. Boxplots, in general, can show symmetry and skewness, but not features of shape like bimodality or uniformity.

c) Median is the better measure of center, since the distribution of ages has outliers. Median is more resistant to outliers than the mean.

d) IQR is a better measure of spread, since the distribution of ages has outliers. IQR is more resistant to outliers than the standard deviation.

23. Graduation?

a) The distribution of the percent of incoming college freshman who graduate on time is roughly symmetric. The mean and the median are reasonably close to one another and the quartiles are approximately the same distance from the mean.

b) Upper Fence:
$$Q3 + 1.5(IQR) = 74.75 + 1.5(74.75 - 59.15)$$
$$= 74.75 + 23.4$$
$$= 98.15$$

Lower Fence:
$$Q1 - 1.5(IQR) = 59.15 - 1.5(74.75 - 59.15)$$
$$= 59.15 - 23.4$$
$$= 35.75$$

Since the maximum value of the distribution of the percent of incoming freshmen who graduate on time is 87.4% and the upper fence is 98.15%, there are no high outliers. Likewise, since the minimum is 43.2% and the lower fence is 35.75%, there are no low outliers. Since the minimum and maximum percentages are within the fences, all percentages must be within the fences.

c) A boxplot of the distribution of the percent of incoming freshmen who graduate on time is at the right.

d) The distribution of the percent of incoming freshmen who graduate on time is roughly symmetric, with mean of approximately 68% of freshmen graduating on time. Universities surveyed had between 43.2% and 87.4% of students graduating on time, with the middle 50% of universities reporting between 59.15% and 74.75% graduating on time.

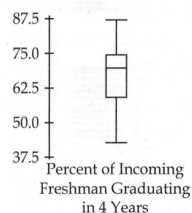

Percent of Incoming Freshman Graduating in 4 Years

25. Caffeine.

a) *Who* – 45 volunteeers. *What* – Level of caffeine consumption and memory test score. *When* – Not specified. *Where* – Not specified. *Why* – The student researchers want to see the possible effects of caffeine on memory. *How* – It appears that the researchers imposed the treatment of level of caffeine consumption in an experiment. However, this point is not clear. Perhaps they allowed the subjects to choose their own level of caffeine.

b) *Variables* – Caffeine level is a categorical variable with three levels: no caffeine, low caffeine, and high caffeine. Test score is a quantitative variable, measured in number of items recalled correctly.

c)

d) The groups consuming no caffeine and low caffeine had comparable memory test scores. A typical score from these groups was around 21. However, the scores of the group consuming no caffeine were more consistent, with a smaller range and smaller interquartile range than the scores of the group consuming low caffeine. The group consuming high caffeine had lower memory scores in general, with a median score of about 19. No one in the high caffeine group scored above 24, but 25% of each of the other groups scored above 24.

27. States.

a) The distribution of state populations is skewed heavily to the right. Therefore, the median and IQR are the appropriate measures of center and spread.

b) The mean population must be larger than the median population. The extreme values on the right affect the mean greatly and have no effect on the median.

c) There are 51 entries in the stemplot, so the 26th entry must be the median. Counting in the ordered stemplot gives median = 4 million people. The middle of the lower 50% of the list (26 state populations) is between the 13th and 14th population, or 1.5 million people. The middle of the upper half of the list (26 state populations) is between the 13th and 14th population from the top, or 6 million people. The IQR = Q3 – Q1 = 6 – 1.5 = 4.5 million people.

d) The distribution of population for the 50 U.S. States and Washington, D.C. is skewed heavily to the right. The median population is 4 million people, with 50% of states having populations between 1 and 6 million people. There is one outlier, a state with 34 million people. The next highest population is only 21 million.

29. Derby speeds.

a) The median speed is the speed at which 50% of the winning horses ran slower. Find 50% on the left, move straight over to the graph and down to a speed of about 36 mph.

b) Quartile 1 is at 25% on the left, and Quartile 3 is at 75% on the left. Matching these to the ogive, Q1 = 34.5 mph and Q3 = 36.5 mph.

c) Range = Max – Min = 38 – 31 = 7 mph
IQR = Q3 – Q1 = 36.5 – 34.5 = 2 mph

d) A boxplot of winning Kentucky Derby Speeds is at the right.

Derby Speeds

e) The distribution of winning speeds in the Kentucky Derby is skewed to the left. The lowest winning speed is just under 31 mph and the fastest speed is about 37.5 mph. The median speed is approximately 36 mph, and 75% of winning speeds are above 34.5 mph. Only a few percent of winners have had speeds below 33 mph.

31. Reading scores.

a) The highest score for boys was 6, which is higher than the highest score for girls, 5.9.

b) The range of scores for boys is greater than the range of scores for girls.
Range = Max – Min   Range(Boys) = 4   Range(Girls) = 3.1

c) The girls had the greater IQR.
IQR = Q3 – Q1   IQR(Boys) = 4.9 – 3.9 = 1   IQR(Girls) = 5.2 – 3.8 = 1.4

d) The distribution of boys' scores is more skewed. The quartiles are not the same distance from the median. In the distribution of girls' scores, Q1 is 0.7 units below the median, while Q3 is 0.7 units above the median.

e) Overall, the girls did better on the reading test. The median, 4.5, was higher than the median for the boys, 4.3. Additionally, the upper quartile score was higher for girls than boys, 5.2 compared to 4.9. The girls' lower quartile score was slightly lower than the boys' lower quartile score, 3.8 compared to 3.9.

f) The overall mean is calculated by weighting each mean by the number of students.

33. Phone calls.

a) A boxplot of the distribution of Net2Phone rates is at the right.

Net2Phone Rates

b) The mean rate for Net2Phone is 13.7 cents per minute. The median rate is 9.9 cents per minute. The rates for India and Pakistan are outliers, so the median is the appropriate measure of center.

c) The IQR of the distribution of Net2Phone rates is 7.6 cents per minute and the standard deviation is 11.8 cents per minute. The rates for India and Pakistan are outliers, so the IQR is the appropriate measure of spread.

d) Upper Fence: $Q3 + 1.5(IQR) = 15.5 + 1.5(15.5 - 7.9)$
$$= 15.5 + 11.4$$
$$= 26.9$$

Lower Fence: $Q1 - 1.5(IQR) = 7.9 - 1.5(15.5 - 7.9)$
$$= 7.9 - 11.4$$
$$= -3.5$$

Any country that has a rate higher than 26.9 cents per minute is an outlier. Both India and Pakistan have rates of 49 cents per minute. No country could be a low outlier. It is impossible to have a rate lower than –3.5 cents per minute.

e) The distribution of the long distance rates for Net2Phone is unimodal and symmetric except for two outliers at 49 cents per minute. These are the rates for India and Pakistan. The median rate is 9.9 cents per minute. The middle 50% of rates are between 7.9 and 15.5 cents per minute for an IQR of 7.6 cents per minute.

f) The rates of these 24 countries may not be representative of the rates of all 250 countries that are serviced by Net2Phone. It would be unwise to assume that these rates could tell us anything about Net2Phone rates in general.

35. Math scores.

a) 5-number summary: 275, 448, 499, 531, 604
IQR: 83
Mean: 487.2
Standard deviation: 72.3

b)

Math Scores

Math Scores

The distribution of average math achievement scores for eighth graders in 38 nations is unimodal, skewed slightly to the left, and centered at about 500. South Africa's score of 275 is the lowest and is a departure from the overall pattern of the data. Not including this score, scores range from 337 to 604, with the middle 50% of the scores falling between 448 and 531. The United States score of 502, although well above the mean of 487.2, is actually near the middle of the distribution. Several low scores pull down the mean, making the median a better measure of center.

37. Gasoline usage.

In 2000, per capita gasoline usage by state in the United States averaged approximately 500 gallons. The distribution of gasoline usage was slightly skewed to the left, with two low outliers, North Carolina and Hawaii. These states used much less gasoline per person than other states. The IQR of the distribution was 82 gallons per person, with the middle 50% of states having gasoline usage between 453 and 535 gallons per person.

Gasoline Usage by U.S. State 1999

39. Customer Database.

a) The mean of 54.41 is meaningless. The codes assigned to the titles are categories, even though the categories are represented by numbers. Averaging is only appropriate for quantitative data.

b) Typically, the mean and standard deviation are influenced by outliers and skewness.

c) There is no point in discussing the reason for the differences in the summary statistics, because the summary statistics are only appropriate for quantitative data. The title codes are categorical.

41. Eye and hair color.

The graph is not appropriate. Boxplots are for quantitative data, and these are categorical data, although coded as numbers. The numbers used for hair color and eye color are arbitrary, so the boxplot and any accompanying statistics for eye color make no sense.

43. Stereograms, revisited.

The re-expression using logarithms has a distribution that is more symmetric than the original distribution of fusion times, and the re-expression has no outliers. This symmetry makes it easier to compare the two groups.

Chapter 6 – The Standard Deviation as a Ruler and the Normal Model

1. **Payroll.**

   a) The distribution of salaries in the company's weekly payroll is skewed to the right. The mean salary, $700, is higher than the median, $500.

   b) The IQR, $600, measures the spread of the middle 50% of the distribution of salaries.

   $$Q3 - Q1 = IQR$$
   $$Q3 = Q1 + IQR$$
   $$Q3 = \$350 + \$600$$
   $$Q3 = \$950$$

   50% of the salaries are found between $350 and $950.

   c) If a $50 raise were given to each employee, all measures of center or position would increase by $50. The minimum would change to $350, the mean would change to $750, the median would change to $550, and the first quartile would change to $400. Measures of spread would not change. The entire distribution is simply shifted up $50. The range would remain at $1200, the IQR would remain at $600, and the standard deviation would remain at $400.

   d) If a 10% raise were given to each employee, all measures of center, position, and spread would increase by 10%.

Minimum = $330	Mean = $770	Median = $550	Range = $1320
IQR = $660	First Quartile = $385	St. Dev. = $440	

3. **SAT or ACT?**

   Measures of center and position (lowest score, top 25% above, mean, and median) will be multiplied by 40 and increased by 150 in the conversion from ACT to SAT by the rule of thumb. Measures of spread (standard deviation and IQR) will only be affected by the multiplication.

Lowest score = 910	Mean = 1230	Standard deviation = 120
Top 25% above = 1350	Median = 1270	IQR = 240

5. **Temperatures.**

   In January, with mean temperature 36° and standard deviation in temperature 10°, a high temperature of 55° is almost 2 standard deviations above the mean. In July, with mean temperature 74° and standard deviation 8°, a high temperature of 55° is more than two standard deviations below the mean. A high temperature of 55° is less likely to happen in July, when 55° is farther away from the mean.

7. **Final exams.**

   a) Anna's average is $\dfrac{83 + 83}{2} = 83$. Megan's average is $\dfrac{77 + 95}{2} = 86$.

   Only Megan qualifies for language honors, with an average higher than 85.

b) On the French exam, the mean was 81 and the standard deviation was 5. Anna's score of 83 was 2 points, or 0.4 standard deviations, above the mean. Megan's score of 77 was 4 points, or 0.8 standard deviations below the mean.

On the Spanish exam, the mean was 74 and the standard deviation was 15. Anna's score of 83 was 9 points, or 0.6 standard deviations, above the mean. Megan's score of 95 was 21 points, or 1.4 standard deviations, above the mean.

Measuring their performance in standard deviations is the only fair way in which to compare the performance of the two women on the test.

Anna scored 0.4 standard deviations above the mean in French and 0.6 standard deviations above the mean in Spanish, for a total of 1.0 standard deviation above the mean.

Megan scored 0.8 standard deviations below the mean in French and 1.4 standard deviations above the mean in Spanish, for a total of only 0.6 standard deviations above the mean.

Anna did better overall, but Megan had the higher average. This is because Megan did very well on the test with the higher standard deviation, where it was comparatively easy to do well.

9. **Professors.**

The standard deviation of the distribution of years of teaching experience for college professors must be 6 years. College professors can have between 0 and 40 (or possibly 50) years of experience. A workable standard deviation would cover most of that range of values with ±3 standard deviations around the mean. If the standard deviation were 6 months ($\frac{1}{2}$ year), some professors would have years of experience 10 or 20 standard deviations away from the mean, whatever it is. That isn't possible. If the standard deviation were 16 years, ±2 standard deviations would be a range of 64 years. That's way too high. The only reasonable choice is a standard deviation of 6 years in the distribution of years of experience.

11. **Guzzlers.**

a) The Normal model for auto fuel economy is at the right.

b) Approximately 68% of the cars are expected to have highway fuel economy between 18.6 mpg and 31.0 mpg.

c) Approximately 16% of the cars are expected to have highway fuel economy above 31 mpg.

d) Approximately 13.5% of the cars are expected to have highway fuel economy between 31 mpg and 37 mpg.

e) The worst 2.5% of cars are expected to have fuel economy below approximately 12.4 mpg.

13. Winter Olympics 2002 downhill.

a) The 2002 Winter Olympics downhill times have mean of 102.71 second and standard deviation 3.01 seconds. 99.7 seconds is 1 standard deviation below the mean. If the Normal model is appropriate, 16% of the times should be below 99.7 seconds.

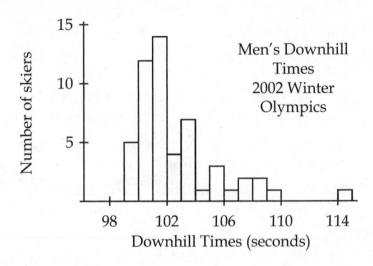

b) Only 3 out of 53 times (5.7%) are below 99.7 seconds.

c) The percentages in parts a and b do not agree because the Normal model is not appropriate in this situation.

d) The histogram of 2002 Winter Olympic Downhill times is skewed to the right, and has a high outlier. The Normal model is not appropriate for the distribution of times, because the distribution is not unimodal and symmetric.

15. Trees.

a) The Normal model for the distribution of tree diameters is at the right.

b) Approximately 95% of the trees are expected to have diameters between 1.0 inch and 19.8 inches.

c) Approximately 2.5% of the trees are expected to have diameters less than an inch.

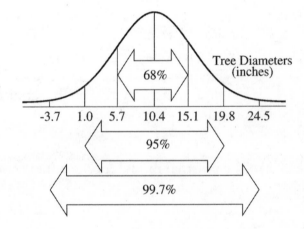

d) Approximately 34% of the trees are expected to have diameters between 5.7 inches and 10.4 inches.

e) Approximately 16% of the trees are expected to have diameters over 15 inches.

17. TV watching.

a) Approximately 16% of the college students are expected to watch less than 1 standard deviation below the mean number of hours of TV.

b) The distribution of the number of hours of TV watched per week has mean 3.66 hours and standard deviation 4.93 hours. According to the Normal model, students who watch fewer than 1 standard deviation below the mean number of hours of TV are expected to watch less than –1.27 hours of TV per week. Of course, it is impossible to watch less than 0 hours of TV, let alone less than –1.27 hours.

c) The distribution of the number of hours of TV watched per week by college students is skewed heavily to the right. Use of the Normal model is not appropriate for this distribution, since it is not unimodal and symmetric.

19. Normal models.

a)

6.68%

$z = 1.5$

b)

98.78%

$z = 2.25$

c)

71.63%

$z = -1$ $z = 1.15$

d)

61.71%

$z = -0.5$ $z = 0.5$

21. More Normal models.

a)

20%

$z = 0.842$

b)

75%

$z = -0.674$

c)

3%

$z = -1.88$

d)

90%

$z = -1.645$ $z = 1.645$

23. Parameters.

a)

$$z = \frac{y - \mu}{\sigma}$$

$$0.126 = \frac{30 - 20}{\sigma}$$

$$0.126\sigma = 10$$

$$\sigma = 79.58$$

b)

$$z = \frac{y - \mu}{\sigma}$$

$$-2.054 = \frac{50 - 88}{\sigma}$$

$$-2.054\sigma = -38$$

$$\sigma = 18.50$$

c)

$$z = \frac{y - \mu}{\sigma}$$

$$0.842 = \frac{100 - \mu}{5}$$

$$(0.842)(5) = 100 - \mu$$

$$\mu = 95.79$$

d)

$$z = \frac{y - \mu}{\sigma}$$

$$1.282 = \frac{17.2 - \mu}{15.6}$$

$$(1.282)(15.6) = 17.2 - \mu$$

$$\mu = -2.79$$

25. Cholesterol.

a) The Normal model for cholesterol levels of adult American women is at the right.

b)

$$z = \frac{y - \mu}{\sigma}$$

$$z = \frac{200 - 188}{24}$$

$$z = 0.5$$

According to the Normal model, 30.85% of American women are expected to have cholesterol levels over 200.

c)

According to the Normal model, 17.00% of American women are expected to have cholesterol levels between 150 and 170.

d)

According to the Normal model, the interquartile range of the distribution of cholesterol levels of American women is approximately 32.38 points.

e)

$$z = \frac{y - \mu}{\sigma}$$

$$1.036 = \frac{y - 188}{24}$$

$$y = 212.87$$

According to the Normal model, the highest 15% of women's cholesterol levels are above approximately 212.87 points.

27. Kindergarten.

a)

$$z = \frac{y - \mu}{\sigma}$$

$$z = \frac{36 - 38.2}{1.8}$$

$$z = -1.222$$

According to the Normal model, approximately 11.1% of kindergarten kids are expected to be less than three feet (36 inches) tall.

b)

$$z = \frac{y - \mu}{\sigma}$$

$$-1.282 = \frac{y_1 - 38.2}{1.8}$$

$$y_1 = 35.89$$

$$z = \frac{y - \mu}{\sigma}$$

$$1.282 = \frac{y_2 - 38.2}{1.8}$$

$$y_2 = 40.51$$

According to the Normal model, the middle 80% of kindergarten kids are expected to be between 35.89 and 40.51 inches tall. (The appropriate values of $z = \pm 1.282$ are found by using right and left tail percentages of 10% of the Normal model.)

c)

$$z = \frac{y - \mu}{\sigma}$$

$$1.282 = \frac{y - 38.2}{1.8}$$

$$y = 40.51$$

According to the Normal model, the tallest 10% of kindergarteners are expected to be at least 40.51 inches tall.

29. First steps.

A good way to visualize the solution to this problem is to look at the distance between 10 and 13 months in two different scales. First, 10 and 13 months are 3 months apart. When measured in standard deviations, the respective *z*-scores, -1.645 and 0.674, are 2.319 standard deviations apart. So, 3 months must be the same as 2.319 standard deviations.

According to the Normal model, the mean age at which babies develop the ability to walk is 12.1 months, with a standard deviation of 1.3 months.

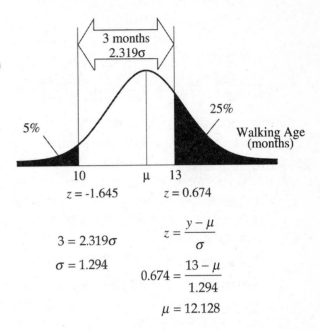

$$3 = 2.319\sigma$$

$$\sigma = 1.294$$

$$z = \frac{y - \mu}{\sigma}$$

$$0.674 = \frac{13 - \mu}{1.294}$$

$$\mu = 12.128$$

31. Eggs.

a)

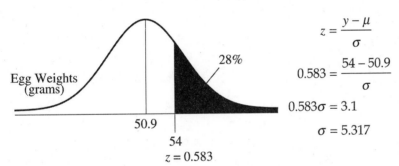

$$z = \frac{y - \mu}{\sigma}$$

$$0.583 = \frac{54 - 50.9}{\sigma}$$

$$0.583\sigma = 3.1$$

$$\sigma = 5.317$$

According to the Normal model, the standard deviation of the egg weights for young hens is expected to be 5.3 grams.

b)

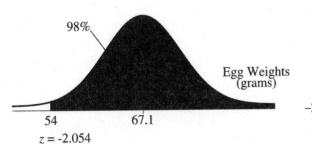

$$z = \frac{y - \mu}{\sigma}$$

$$-2.054 = \frac{54 - 67.1}{\sigma}$$

$$-2.054\sigma = -13.1$$

$$\sigma = 6.377$$

According to the Normal model, the standard deviation of the egg weights for older hens is expected to be 6.4 grams.

c) The younger hens lay eggs that have more consistent weights than the eggs laid by the older hens. The standard deviation of the weights of eggs laid by the younger hens is lower than the standard deviation of the weights of eggs laid by the older hens.

d) A good way to visualize the solution to this problem is to look at the distance between 54 and 70 grams in two different scales. First, 54 and 70 grams are 16 grams apart. When measured in standard deviations, the respective z-scores, -1.405 and 1.175, are 2.580 standard deviations apart. So, 16 grams must be the same as 2.580 standard deviations.

$$16 = 2.58\sigma$$

$$\sigma = 6.202$$

$$z = \frac{y - \mu}{\sigma}$$

$$1.175 = \frac{70 - \mu}{6.202}$$

$$\mu = 62.713$$

According to the Normal model, the mean weight of the eggs is 62.7 grams, with a standard deviation of 6.2 grams.

Review of Part I – Exploring and Understanding Data

1. Bananas.

a) A histogram of the prices of bananas from 15 markets, as reported by the USDA, appears at the right.

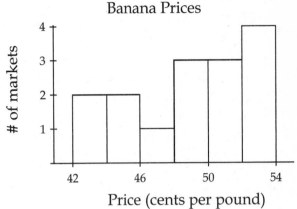

Banana Prices

b) The distribution of banana prices is skewed to the left, so median and IQR are appropriate measures of center and spread.
Median = 49 cents per pound
IQR = 7 cents per pound

c) The distribution of the prices of bananas from 15 markets, as reported by the USDA, is unimodal and skewed to the left. The center of the distribution is approximately 50 cents, with the lowest price 42 cents per pound and the highest price 53 cents per pound.

3. Singers.

The distribution of heights of each voice part is roughly symmetric. The basses and tenors are generally taller than the altos and sopranos, with the basses being slightly taller than the tenors. The sopranos and altos have about the same median height. Heights of basses and sopranos are more consistent than altos and tenors.

5. Beanstalks.

a) The greater standard deviation for the distribution of women's heights means that their heights are more variable than the heights of men.

b) The z-score for women to qualify is 3.25 compared with 4.16 for men, so it is harder for men to qualify. (There appears to be some error in the statistics provided from the National Health Survey. Mean heights of 63.6" (almost 5'4") for men and 60.9" (almost 5'1") for women seem low. The z-scores confirm this skepticism. If the statistics were correct, Beanstalk Clubs would be expected to have virtually no members. According to the Normal model, only 0.0016% of men and 0.058% of women would be expected to qualify for membership!)

7. Watsamatta University.

a) *Who* – Local residents near Watsamatta University. *What* – Age, whether or not the respondent attended college, and whether or not the respondent had a favorable opinion of Watsamatta University. *When* – Not specified. *Where* – Region around Watsamatta University. *Why* – The information will be included in a report to the University's directors. *How* – 850 local residents were surveyed by phone.

b) There is one quantitative variable, age, probably measured in years. There are two categorical variables, college attendance (yes or no), and opinion of Watsamatta University (favorable or unfavorable).

c) There are several problems with the design of the survey. No mention is made of a random selection of residents. Furthermore, there may be a non-response bias present. People with an unfavorable opinion of the university may hang up as soon as the staff member identifies himself or herself. Also, response bias may be introduced by the interviewer. The responses of the residents may be influenced by the fact that employees of the university are asking the questions. There may be greater percentage of favorable responses to the survey than truly exist.

9. Fraud detection.

a) Even though they are numbers, the SIC code is a categorical variable. A histogram is a quantitative display, so it is not appropriate.

b) The Normal model will not work at all. The Normal model is for modeling distributions of unimodal and symmetric quantitative variables. SIC code is a categorical variable.

11. Cramming.

a) Comparitive boxplots of the distributions of Friday and Monday scores are at the right.

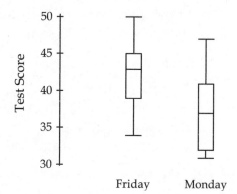

b) The distribution of scores on Friday was generally higher by about 5 points. Students fared worse on Monday after preparing for the test on Friday. The spreads are about the same, but the scores on Monday are slightly skewed to the right.

c) A histogram of the distribution of change in test score is at the right.

d) The distribution of changes in score is roughly unimodal and symmetric, and is centered near 4 points. Changes ranged from a student who scored 5 points higher on Monday, to two students who each scored 14 points higher on Friday. Only three students did better on Monday.

13. Let's play cards.

a) Suit is a categorical variable.

b) In the game of Go Fish, the denomination is not ordered. Numbers are merely matched with one another. You may have seen children's Go Fish decks that have symbols or pictures on the cards instead of numbers. These work just fine.

c) In the game of Gin Rummy, the order of the cards is important. During the game, ordered "runs" of cards are assembled (with Jack = 11, Queen = 12, King = 13), and at the end of the hand, points are totaled from the denomination of the card (face cards = 10 points). However, even in Gin Rummy, the denomination of the card sometimes behaves like a categorical variable. When you are collecting 3s, for example, order doesn't matter.

15. Hard water.

a) The variables in this study are both quantitative. Annual mortality rate for males is measured in deaths per 100,000. Calcium concentration is measured in parts per million.

b) The distribution of calcium concentration is skewed right, with many towns having concentrations below 25 ppm. The rest of the towns have calcium concentrations which are distributed in a fairly uniform pattern from 25 ppm to 100 ppm, tapering off to a maximum concentration around 150 ppm.
The distribution of mortality rates is unimodal and symmetric, with center approximately 1500 deaths per 100,000. The distribution has a range of 1000 deaths per 100,000, from 1000 to 2000 deaths per 100,000.

17. Seasons.

a) The two histograms have different horizontal and vertical scales. This makes a quick comparison impossible.

b) The center of the distribution of average temperatures in January in is the low 30s, compared to a center of the distribution of July temperatures in the low 70s. The January distribution is also much more spread out than the July distribution. The range is over 50 degrees in January, compared to a range of over 20 in July. The distribution of average temperature in January is skewed slightly to the right, while the distribution of average temperature in July is roughly symmetric.

c) The distribution of difference in average temperature (July – January) for 60 large U.S. cities is slightly skewed to the left, with median at approximately 44 degrees. There are several low outliers, cities with very little difference between their average July and January temperatures. The single high outlier is a city with a large difference in average temperature between July and January. The middle 50% of differences are between approximately 38 and 46 degrees.

19. Old Faithful?

a) The distribution of duration of the 222 eruptions is bimodal, with modes at approximately 2 minutes and 4.5 minutes. The distribution is fairly symmetric around each mode.

b) The bimodal shape of the distribution of duration of the 222 eruptions suggests that there may be two distinct groups of eruption durations. Summary statistics would try to summarize these two groups as a single group, which wouldn't make sense.

c) The intervals between eruptions are generally longer for long eruptions than the intervals for short eruptions. Over 75% of the short eruptions had intervals of approximately 60 minutes or less, while almost all of the long eruptions had intervals of more than 60 minutes.

21. Liberty's nose.

a) The distribution of the ratio of arm length to nose length of 18 girls in a statistics class is unimodal and roughly symmetric, with center around 15. There is one low outlier, a ratio of 11.8. A boxplot is provided at the right. A histogram or stemplot is also an appropriate display.

b) In the presence of an outlier, the 5-number summary is the appropriate choice for summary statistics. The 5-number summary is 11.8, 14.4, 15.25, 15.7, 16.9. The IQR is 1.3.

c) The ratio of 9.3 for the Statue of Liberty is very low, well below the lowest ratio in the statistics class, 11.8, which is already a low outlier. Compared to the girls in the statistics class, the Statue of Liberty's nose is very long in relation to her arm.

23. Sample.

Overall, the follow-up group was insured only 11.1% of the time as compared to 16.6% for the not traced group. At first, it appears that group is associated with presence of health insurance. But for blacks, the follow-up group was quite close (actually slightly higher) in terms of being insured: 8.9% to 8.7%. The same is true for whites. The follow-up group was insured 83.3% of the time, compared to 82.5% of the not traced group. When broken down by race, we see that group is not associated with presence of health insurance for either race. This demonstrates Simpson's paradox, because the overall percentages lead us to believe that there is an association between health insurance and group, but we see the truth when we examine the situation more carefully.

25. Be quick!

a) The Normal model for the distribution of reaction times is at the right.

b) The distribution of reaction times is unimodal and symmetric, with mean 1.5 seconds, and standard deviation 0.18 seconds. According to the Normal model, 95% of drivers are expected to have reaction times between 1.14 seconds and 1.86 seconds.

c)

$$z = \frac{y - \mu}{\sigma}$$

$$z = \frac{1.25 - 1.50}{0.18}$$

$$z = -1.389$$

According to the Normal model, 8.24% of drivers are expected to have reaction times below 1.25 seconds.

d)

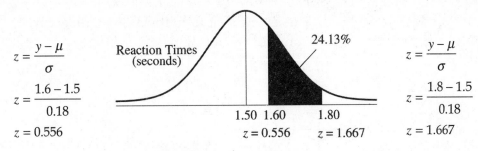

$$z = \frac{y - \mu}{\sigma}$$

$$z = \frac{1.6 - 1.5}{0.18}$$

$$z = 0.556$$

$$z = \frac{y - \mu}{\sigma}$$

$$z = \frac{1.8 - 1.5}{0.18}$$

$$z = 1.667$$

According to the Normal model, 24.13% of drivers are expected to have reaction times between 1.6 seconds and 1.8 seconds.

e)

$$z = \frac{y - \mu.}{\sigma}$$

$$-0.674 = \frac{Q1 - 1.5}{0.18}$$

$$Q1 = 1.38$$

$$z = \frac{y - \mu}{\sigma}$$

$$0.674 = \frac{Q3 - 1.5}{0.18}$$

$$Q3 = 1.62$$

IQR = $(Q3 - Q1)$ = 0.24 seconds

According to the Normal model, the interquartile range of the distribution of reaction times is expected to be 0.24 seconds.

f)

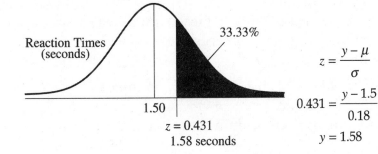

$$z = \frac{y - \mu}{\sigma}$$

$$0.431 = \frac{y - 1.5}{0.18}$$

$$y = 1.58$$

According to the Normal model, the slowest 1/3 of all drivers are expected to have reaction times of 1.58 seconds or more. (Remember that a high reaction time is a SLOW reaction time!)

27. Wines.

a) A histogram of the case prices for 36 Finger Lakes region wines is at the right.

b) Since the distribution of case prices is unimodal and symmetric, the mean and standard deviation are appropriate measures of center and spread. The mean case price is $100.25, and the standard deviation is $25.54.

c) The distribution of case prices of Finger Lakes wines is unimodal and symmetric, with mean $100.25 and standard deviation $25.54. The cheapest case was $52 and the most expensive was $151.

d) 23 of the 36 wines (64%) had case prices within one standard deviation of the mean, or within the interval $74.71 - $125.79. This is fairly close to the 68% predicted by the Normal model. The Normal model may be useful for modeling the distribution of Finger Lakes wine case prices.

29. Engines.

a) The count of cars is 38.

b) The mean displacement is higher than the median displacement, indicating a distribution of displacements that is skewed to the right. There are likely to be several very large engines in a group that consists of mainly smaller engines.

c) Since the distribution is skewed, the median and IQR are useful measures of center and spread. The median displacement is 148.5 cubic inches and the IQR is 126 cubic inches.

d) Your neighbor's car has an engine that is bigger than the median engine, but 227 cubic inches is smaller than the third quartile of 231, meaning that at least 25% of cars have a bigger engine than your neighbor's car. Don't be impressed!

e) Using the Outlier Rule (more than 1.5 IQRs beyond the quartiles) to find the fences:

Upper Fence: Q3 + 1.5(IQR) = 231 + 1.5(126) = 420 cubic inches.

Lower Fence: Q1 – 1.5(IQR) = 105 – 1.5(126) = - 84 cubic inches.

Since there are certainly no engines with negative displacements, there are no low outliers. Q1 + Range = 105 + 275 = 380 cubic inches. This means that the maximum must be less than 380 cubic inches. Therefore, there are no high outliers (engines over 420 cubic inches).

f) It is not reasonable to expect 68% of the car engines to measure within one standard deviation of the mean. The distribution engine displacements is skewed to the right, so the Normal model is not appropriate.

g) Multiplying each of the engine displacements by 16.4 to convert cubic inches to cubic centimeters would affect measures of position and spread. All of the summary statistics (except the count!) could be converted to cubic centimeters by multiplying each by 16.4.

31. Age and party.

a) 1101 of 4002, or approximately 27.5%, of all voters surveyed were Republicans.

b) This was a representative telephone survey conducted by Gallup, a reputable polling firm. It is likely to be a reasonable estimate of the percentage of all voters who are Republicans.

c) 1001 + 1004 = 2005 of 4002, or approximately 50.1%, of all voters surveyed were under 30 or over 65 years old.

d) 409 of 4002, or approximately 10.2%, of all voters surveyed were Independents under the age of 30.

e) 409 of the 1497 Independents surveyed, or approximately 27.3%, were under the age of 30.

f) 409 of the 1001 respondents under 30, or approximately 40.9%, were Independents.

33. Herbal medicine.

a) *Who* – 100 customers. *What* – Researchers asked whether or not the customer had taken the cold remedy and had customers rate the effectiveness of the remedy on a scale from 1 to 10. *When* – Not specified. *Where* – Store where natural health products are sold. *Why* – The researchers were from the Herbal Medicine Council, which sounds suspiciously like a group that might be promoting the use of herbal remedies. *How* – Researchers conducted personal interviews with 100 customers. No mention was made of any type of random selection.

b) "Have you taken the cold remedy?" is a categorical variable. Effectiveness on a scale of 1 to 10 is a categorical variable, as well, with respondents rating the remedy by placing it into one of 10 categories.

c) Very little confidence can be placed in the Council's conclusions. Respondents were people who already shopped in a store that sold natural remedies. They may be pre-disposed to thinking that the remedy was effective. Furthermore, no attempt was made to randomly select respondents in a representative manner. Finally, the Herbal Medicine Council has an interest in the success of the remedy.

35. Bike safety.

a) *Who* – Years from 1991 to 2000. *What* – Number of bicycle injuries reported. *When* – 1991 to 2000. *Where* – Massachusetts. *Why* – The information was collected for a report by the Governor's Highway Safety Bureau. *How* – Although not specifically stated, the information was probably collected from government agency or hospital records.

b)   c)

Massachusetts Yearly
Bicycle Injuries (1991 - 2000)

```
17 | 6
16 |
15 | 2
14 | 5
13 | 1478
12 | 8
11 | 2
10 | 3
```
15|2 = 1515 - 1524
Bicycle Injuries

d) The stem-and-leaf display of the number of yearly bicycle injuries reported in Massachusetts shows that the center of the distribution is approximately 1350 bicycle injuries per year. Additionally, we can see that the year in which 1763 injuries were reported had an unusually high number of injuries. This is not visible on the timeplot.

e) The timeplot of the number of yearly bicycle injuries reported in Massachusetts shows that the number of injuries per year has declined over time.

f) In the years 1991 – 2000, the number of reported bicycle injuries per year has declined from approximately 1800 injuries in 1991 to approximately 1100 injuries in 2000. The number of injuries was unusually high in 1991. In the middle of the decade, the number of bicycle injuries reported was fairly steady at approximately 1300 injuries.

37. Some assembly required.

a) A good way to visualize the solution to this problem is to look at the distance between 1 and 2 hours in two different scales. First, 1 and 2 hours are 1 hour apart. When measured in standard deviations, the respective z-scores, -0.674 and 1.645, are 2.319 standard deviations apart. So, 1 hour must be the same as 2.319 standard deviations.

According to the Normal model, the mean assembly time is 1.29 hours and the standard deviation is 0.43 hours.

$$1 = 2.319\sigma$$

$$\sigma = 0.431$$

$$z = \frac{y - \mu}{\sigma}$$

$$1.645 = \frac{2 - \mu}{0.431}$$

$$\mu = 1.291$$

b)

$$z = \frac{y - \mu}{\sigma}$$

$$0.253 = \frac{y - 1.29}{0.43}$$

$$y = 1.40$$

According to the Normal model, the company would need to claim that the desk takes "less than 1.40 hours to assemble", not the catchiest of slogans!

c)

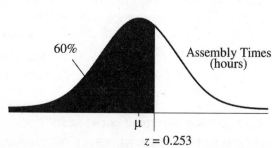

$$z = \frac{y - \mu}{\sigma}$$

$$0.253 = \frac{1 - \mu}{0.43}$$

$$\mu = 0.89$$

According to the Normal model, the company would have to lower the mean assembly time to 0.89 hour (53.4 minutes).

d) The new instructions and part-labeling may have helped lower the mean, but it also may have changed the standard deviation, making the assembly times more consistent as well as lower.

Chapter 7 – Scatterplots, Association, and Correlation

1. **Association.**

   a) Either weight in grams or weight in ounces could be the explanatory or response variable. Greater weights in grams correspond with greater weights in ounces. The association between weight of apples in grams and weight of apples in ounces would be positive, linear, and perfect. Each apple's weight would simply be measured in two different scales. The points would line up perfectly.

   b) Circumference is the explanatory variable, and weight is the response variable, since one-dimensional circumference explains three-dimensional volume (and therefore weight). For apples of roughly the same size, the association would be positive, linear, and strong. If the sample of apples contained very small and very large apples, the association's true curved form would become apparent.

   c) There would be no association between shoe size and GPA of college freshmen.

   d) Number of miles driven is the explanatory variable, and gallons remaining in the tank is the response variable. The greater the number of miles driven, the less gasoline there is in the tank. If a sample of different cars is used, the association is negative, linear, and moderate. If the data is gathered on different trips with the same car, the association would be strong.

3. **Association.**

   a) Altitude is the explanatory variable, and temperature is the response variable. As you climb higher, the temperature drops. The association is negative, linear, and strong.

   b) At first, it appears that there should be no association between ice cream sales and air conditioner sales. When the lurking variable of temperature is considered, the association becomes more apparent. When the temperature is high, ice cream sales tend to increase. Also, when the temperature is high, air conditioner sales tend to increase. Therefore, there is likely to be an increase in the sales of air conditioners whenever there is an increase in the sales of ice cream. The association is positive, linear, and moderate. Either one of the variables could be used as the explanatory variable.

   c) Age is the explanatory variable, and grip strength is the response variable. The association is neither negative nor positive, but is curved, and moderate in scatter, due to the variability in grip strength among people in general. The very young would have low grip strength, and grip strength would increase as age increased. After reaching a maximum (at whatever age physical prowess peaks), grip strength would decline again, with the elderly having low grip strengths.

   d) Blood alcohol content is the explanatory variable, and reaction time is the response variable. As blood alcohol level increase, so does the time it takes to react to a stimulus. The association is positive, probably curved, and strong. The scatterplot would probably be almost linear for low concentrations of alcohol in the blood, and then begin to rise dramatically, with longer and longer reaction times for each incremental increase in blood alcohol content.

5. **Scatterplots.**

   a) None of the scatterplots show little or no association, although # 4 is very weak.

   b) #3 and #4 show negative association. Increases in one variable are generally related to decreases in the other variable.

   c) #2, #3, and #4 each show a linear association.

   d) #2 shows a moderately strong association.

   e) #1 and #3 each show a very strong association. #1 shows a curved association and #3 shows a linear association.

7. **Performance IQ scores *vs.* brain size.**

   The scatterplot of IQ scores *vs.* Brain Sizes is scattered, with no apparent pattern. There appears to be little or no association between the IQ scores and brain sizes displayed in this scatterplot.

9. **Firing pottery.**

   a) A histogram of the number of broken pieces is at the right.

   b) The distribution of the number broken pieces per batch of pottery is skewed right, centered around 1 broken piece per batch. Batches had from 0 and 6 broken pieces. The scatterplot does not show the center or skewness of the distribution.

   c) The scatterplot shows that the number of broken pieces increases as the batch number increases. If the 8 daily batches are numbered sequentially, this indicates that batches fired later in the day generally have more broken pieces. This information is not visible in the histogram.

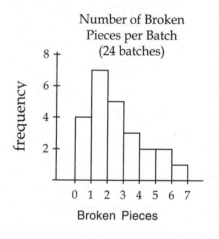

11. **Matching.**

    a) 0.006   b) 0.777   c) - 0.923   d) - 0.487

13. **Lunchtime.**

    a) The correlation between time toddlers spent at the table and the number of calories consumed by the toddlers is $r = -0.649$.

    b) If time spent at the table were recorded in hours instead of minutes, the correlation would not change at all. Correlation is calculated from z-scores, and since these unitless measures of position relative to the mean are unaffected by changes in scale, correlation is likewise unaffected.

    c) The correlation between time spent at the table and calorie consumption for toddlers is −0.649, a moderate, negative correlation. Toddlers who spent more time at the table tended to consume fewer calories.

d) The analyst's remark is speculative. There are many possible explanations for the behavior of the toddlers. The data merely show us an association between time spent at the table and calories consumed by toddlers, and association is not the same thing as a cause-and-effect relationship.

15. Fuel economy.

a) A scatterplot of average fuel economy vs. horsepower ratings is at the right.

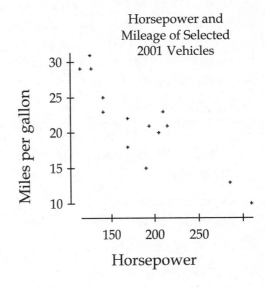

b) There is a strong, negative, linear association between horsepower and mileage of the selected vehicles.

c) The correlation between horsepower and mileage of the selected vehicles is $r = -0.878$.

d) Generally, vehicles in the selected group with more horsepower have lower mileage. There are two vehicles with extremely high horsepower ratings and low mileage. If these two vehicles were removed, the correlation would not be as high.

17. Burgers.

There is no apparent association between the number of grams of fat and the number of milligrams of sodium in several brands of fast food burgers. The correlation is only $r = 0.199$, which is close to zero, an indication of no association. One burger had a much lower fat content than the other burgers, at 19 grams of fat, with 920 milligrams of sodium. Without this (comparatively) low fat burger, the correlation would have been even lower.

19. Attendance.

a) Number of runs scored and attendance are quantitative variables, the relationship between them appears to be linear, and there are no outliers, so calculating a correlation is appropriate.

b) The association between attendance and runs scored is positive, linear, and moderate in strength. Generally, as the number of runs scored increases, so does attendance.

c) There is evidence of an association between attendance and runs scored, but a cause-and-effect relationship between the two is not implied. There may be lurking variables that can account for the increases in each. For example, perhaps winning teams score more runs and also have higher attendance. We don't have any basis to make a claim of causation.

21. Politics.

The candidate might mean that there is an **association** between television watching and crime. The term correlation is reserved for describing linear associations between quantitative variables. We don't know what type of variables "television watching" and "crime" are, but they seem categorical. Even if the variables are quantitative (hours of TV watched per week, and number of crimes committed, for example), we aren't sure that the relationship is linear. The politician also seems to be implying a cause-and-effect relationship between television watching and crime. Association of any kind does not imply causation.

23. Correlation errors.

a) If the association between GDP and infant mortality is linear, a correlation of –0.772 shows a moderate, negative association. Generally, as GDP increases, infant mortality rate decreases.

b) Continent is a categorical variable. Correlation measures the strength of linear associations between quantitative variables.

c) Correlation must be between –1 and 1, inclusive. Correlation can never be 1.22.

d) A correlation, no matter how strong, cannot prove a cause-and-effect relationship.

25. Baldness and heart disease.

Even though the variables baldness and heart disease were assigned numerical values, they are categorical. Correlation is only an appropriate measure of the strength of linear association between quantitative variables. Their conclusion is meaningless.

27. Planets.

a) The association between Position Number of each planet and its distance from the sun (in millions of miles) is very strong, positive and curved. The scatterplot is at the right.

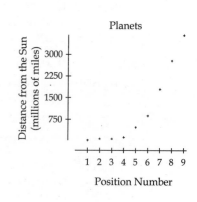

b) The relationship between Position Number and distance from the sun is not linear. Correlation is a measure of the degree of *linear* association between two variables.

c) The scatterplot of the logarithm of distance versus Position Number (shown at the right) still shows a strong, positive relationship, but it is straighter than the previous scatterplot. It still shows a curve in the scatterplot, but it is straight enough that correlation may now be used as an appropriate measure of the strength of the relationship between logarithm of distance and Position Number, which will in turn give an indication of the strength of the association.

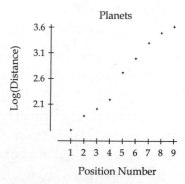

Chapter 8 – Linear Regression

1. Regression equations.

$\bar{x}$	s_x	$\bar{y}$	s_y	r	$\hat{y} = b_0 + b_1 x$
a) 10	2	20	3	0.5	$\hat{y} = 12.5 + .75x$
b) 2	0.06	7.2	1.2	−0.4	$\hat{y} = 23.2 - 8x$
c) 12	6	152	30	−0.8	$\hat{y} = 200 - 4x$
d) 2.5	1.2	25	100	0.6	$\hat{y} = -100 + 50x$

a)

$$b_1 = r\left(\frac{s_y}{s_x}\right)$$
$$b_1 = 0.5\left(\frac{3}{2}\right)$$
$$b_1 = 0.75$$

$$\hat{y} = b_0 + b_1 x$$
$$\bar{y} = b_0 + b_1 \bar{x}$$
$$20 = b_0 + 0.75(10)$$
$$b_0 = 12.5$$

b)

$$b_1 = r\left(\frac{s_y}{s_x}\right)$$
$$b_1 = -0.4\left(\frac{1.2}{0.06}\right)$$
$$b_1 = -8$$

$$\hat{y} = b_0 + b_1 x$$
$$\bar{y} = b_0 + b_1 \bar{x}$$
$$7.2 = b_0 - 8(2)$$
$$b_0 = 23.2$$

c)

$$\hat{y} = b_0 + b_1 x$$
$$\bar{y} = b_0 + b_1 \bar{x}$$
$$\bar{y} = 200 - 4(12)$$
$$\bar{y} = 152$$

$$b_1 = r\left(\frac{s_y}{s_x}\right)$$
$$-4 = -0.8\left(\frac{s_y}{6}\right)$$
$$s_y = 30$$

d)

$$\hat{y} = b_0 + b_1 x$$
$$\bar{y} = b_0 + b_1 \bar{x}$$
$$\bar{y} = -100 + 50(2.5)$$
$$\bar{y} = 25$$

$$b_1 = r\left(\frac{s_y}{s_x}\right)$$
$$50 = r\left(\frac{100}{1.2}\right)$$
$$r = 0.6$$

3. Residuals.

a) The scattered residuals plot indicates an appropriate linear model.

b) The curved pattern in the residuals plot indicates that the linear model is not appropriate. The relationship is not linear.

c) The fanned pattern indicates heteroscedastic data. The models predicting power decreases as the values of the explanatory variable increase.

5. Least squares.

If the 4 x-values are plugged into $\hat{y} = 7 + 1.1x$, the 4 predicted values are $\hat{y} = 18, 29, 51$ and 62, respectively. The 4 residuals are −8, 21, −31, and 18. The squared residuals are 64, 441, 961, and 324, respectively. The sum of the squared residuals is 1790. Least squares means that no other line has a sum lower than 1790. In other words, it's the best fit.

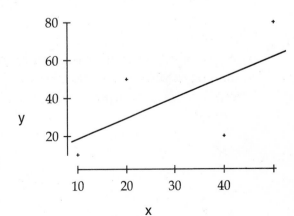

7. **Real estate.**

   a) The explanatory variable (x) is size, measured in square feet, and the response variable (y) is price measured in thousands of dollars. Therefore, the units of the slope are thousands of dollars per square foot.

   b) 71.4% of the variability in price can be explained by variability in size. (In other words, 71.4% of the variability in price can be explained by the linear model.)

   c) The slope of the regression line predicting price from size should be positive. Bigger homes are expected to cost more.

   d) The correlation between size and price is $r = \sqrt{R^2} = \sqrt{0.714} = 0.845$. The positive value of the square root is used, since the relationship is believed to be positive.

   e) The price of a home that is one standard deviation above the mean size would be predicted to be 0.845 standard deviations (in other words r standard deviations) above the mean price.

   f) The price of a home that is two standard deviations below the mean size would be predicted to be 1.69 (or 2×0.845) standard deviations below the mean price.

9. **More real estate.**

   a) According to the linear model, the price of a home is expected to increase $61 (0.061 thousand dollars) for each additional square-foot in size.

   b)
   $$\hat{price} = 47.82 + 0.061x$$
   $$\hat{price} = 47.82 + 0.061(3000)$$
   $$\hat{price} = 230.82$$

   According to the linear model, a 3000 square-foot home is expected to have a price of $230,820.

   c)
   $$\hat{price} = 47.82 + 0.061x$$
   $$\hat{price} = 47.82 + 0.061(1200)$$
   $$\hat{price} = 121.02$$

   According to the linear model, a 1200 square-foot home is expected to have a price of $121,020. The asking price is $121,020 - $6000 = $115,020. $6000 is the (negative) residual.

11. **What slope?**

   The only slope that makes sense is 300 pounds per foot. 30 pounds per foot is too small. For example, a Honda Civic is about 14 feet long, and a Cadillac DeVille is about 17 feet long. If the slope of the regression line was 30 pounds per foot, the Cadillac would be predicted to outweigh the Civic by only 90 pounds! (The real difference is about 1500 pounds.) Similarly, 3 pounds per foot is too small. A slope of 3000 pounds per foot would predict a weight difference of 9000 pounds (4.5 tons) between Civic and DeVille. The only answer that is even reasonable is 300 pounds per foot, which predicts a difference of 900 pounds. This isn't very close to the actual difference of 1500 pounds, but at least it is in the right ballpark.

13. Misinterpretations.

a) R^2 is an indication of the strength of the model, not the appropriateness of the model. A scattered residuals plot is the indicator of an appropriate model.

b) Regression models give predictions, not actual values. The student should have said, "The model predicts that a bird 10 inches tall is expected to have a wingspan of 17 inches."

15. ESP.

a) First, since no one has ESP, you must have scored 2 standard deviations above the mean by chance. On your next attempt, you are unlikely to duplicate the extraordinary event of scoring 2 standard deviations above the mean. You will likely "regress" towards the mean on your second try, getting a lower score. If you want to impress your friend, don't take the test again. Let your friend think you can read his mind!

b) Your friend doesn't have ESP, either. No one does. Your friend will likely "regress" towards the mean score on his second attempt, as well, meaning his score will probably go up. If the goal is to get a higher score, your friend should try again.

17. SAT scores.

a) The association between SAT Math scores and SAT Verbal Scores is linear, moderate in strength, and positive. Students with high SAT Math scores typically have high SAT Verbal scores.

b) One student got a 500 Verbal and 800 Math. That set of scores doesn't seem to fit the pattern.

c) $r = 0.685$ indicates a moderate, positive association between SAT Math and SAT Verbal, but only because the scatterplot shows a linear relationship. Students who scored one standard deviation above the mean in SAT MAth are expected to score 0.685 standard deviations above the mean in SAT Verbal. Additionally, $R^2 = (0.685)^2 = 0.469225$, so 46.9% of the variability in math score is explained by variability in verbal score.

d) The scatterplot of verbal and math scores shows a relationship that is straight enough, so a linear model is appropriate.

$$b_1 = r\left(\frac{s_{Math}}{s_{Verbal}}\right)$$

$$\hat{y} = b_0 + b_1 x$$

$$b_1 = 0.685\left(\frac{96.1}{99.5}\right)$$

$$\bar{y} = b_0 + b_1\bar{x}$$

$$612.2 = b_0 + 0.661593(596.3)$$

$$b_1 = 0.661593$$

$$b_0 = 217.692$$

The equation of the least squares regression line for predicting SAT Math score from SAT Verbal score is $M\hat{a}th = 217.692 + 0.662(Verbal)$.

e) For each additional point in verbal score, the model predicts an increase of 0.662 points in math score. A more meaningful interpretation might be scaled up. For each additional 10 points in verbal score, the model predicts an increase of 6.62 points in math score.

f)

$\widehat{Math} = 217.692 + 0.662(Verbal)$

$\widehat{Math} = 217.692 + 0.662(500)$

$\widehat{Math} = 548.692$

According to the model, a student with a verbal score of 500 is expected to have a math score of 548.692.

g)

$\widehat{Math} = 217.692 + 0.662(Verbal)$

$\widehat{Math} = 217.692 + 0.662(800)$

$\widehat{Math} = 747.292$

According to the model, a student with a verbal score of 800 is expected to have a math score of 747.292. Since she actually scored 800 on math, her residual is 800 – 747.292 = 52.708 points

19. SAT, take 2.

a) $r = 0.685$. The correlation between SAT Math and SAT Verbal is a unitless measure of the degree of linear association between the two variables. It doesn't depend on the order in which you are making predictions.

b) The scatterplot of verbal and math scores shows a relationship that is straight enough, so a linear model is appropriate.

$b_1 = r\left(\dfrac{s_{Verbal}}{s_{Math}}\right)$

$b_1 = 0.685\left(\dfrac{99.5}{96.1}\right)$

$b_1 = 0.709235$

$\hat{y} = b_0 + b_1 x$

$\bar{y} = b_0 + b_1 \bar{x}$

$596.3 = b_0 + 0.709235(612.2)$

$b_0 = 162.106$

The equation of the least squares regression line for predicting SAT Verbal score from SAT Math score is:
$\widehat{Verbal} = 162.106 + 0.709(Math)$

c) A positive residual means that the student's actual verbal score is higher than the score the model predicts for someone with the same math score.

d)

$\widehat{Verbal} = 162.106 + 0.709(Math)$

$\widehat{Verbal} = 162.106 + 0.709(500)$

$\widehat{Verbal} = 516.606$

According to the model, a person with a math score of 500 is expected to have a verbal score of 516.606 points.

e)

$\widehat{Math} = 217.692 + 0.662(Verbal)$

$\widehat{Math} = 217.692 + 0.662(516.606)$

$\widehat{Math} = 559.685$

According to the model, a person with a verbal score of 516.606 is expected to have a math score of 559.685 points.

f) The prediction in part e) does not cycle back to 500 points because the regression equation used to predict math from verbal is a different equation than the regression equation used to predict verbal from math. One was generated by minimizing squared residuals in the verbal direction, the other was generated by minimizing squared residuals in the math direction. If a math score is one standard deviation above the mean, its predicted verbal score regresses toward the mean. The same is true for a verbal score used to predict a math score.

21. Used cars.

a) We are attempting to predict the price in dollars of used Toyota Corollas from their age in years. A scatterplot of the relationship is at the right.

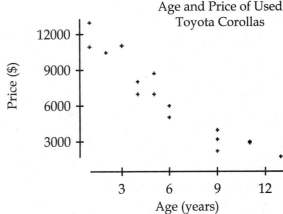

b) There is a strong, negative, linear association between price and age of used Toyota Corollas.

c) The scatterplot provides evidence that the relationship is linear. A linear model will likely be an appropriate model.

d) Since $R^2 = 0.894$, simply take the square root to find r. $\sqrt{0.894} = 0.946$. Since association between age and price is negative, $r = -0.946$.

e) 89.4% of the variability in price of a used Toyota Corolla can be explained by variability in the age of the car.

f) The relationship is not perfect. Other factors, such as options, condition, and mileage explain the rest of the variability in price.

23. More used cars.

a) The scatterplot from Exercise 21 shows that the relationship is straight, so the linear model is appropriate.
The regression equation to predict the price of a used Toyota Corolla from its age is $\hat{Price} = 12319.6 - 924(Years)$.
The computer regression output used is at the right.

Dependent variable is: Price
No Selector
R squared = 89.4%   R squared (adjusted) = 88.7%
s = 1221 with 17 - 2 = 15 degrees of freedom

Source	Sum of Squares	df	Mean Square	F-ratio
Regression	187830720	1	187830720	126
Residual	22346074	15	1489738	

Variable	Coefficient	s.e. of Coeff	t-ratio	prob
Constant	12319.6	575.7	21.4	≤ 0.0001
Age	-924.000	82.29	-11.2	≤ 0.0001

b) According to the model, for each additional year in age, the car is expected to drop $924 in price.

c) The model predicts that a new Toyota Corolla (0 years old) will cost $12,319.60.

d)

$\widehat{Price} = 12319.6 - 924(Years)$

$\widehat{Price} = 12319.6 - 924(7)$

$\widehat{Price} = 5851.60$

According to the model, an appropriate price for a 7-year old Toyota Corolla is $5,851.60.

e) Buy the car with the negative residual. Its actual price is lower than predicted.

f)

$\widehat{Price} = 12319.6 - 924(Years)$

$\widehat{Price} = 12319.6 - 924(10)$

$\widehat{Price} = 3079.60$

According to the model, a 10-year-old Corolla is expected to cost $3,079.60. The car has an actual price of $1500, so its residual is $1500 − $3079.60 = − $1579.60 The car costs $1579.60 less than predicted.

g) The model would not be useful for predicting the price of a 20-year-old Corolla. The oldest car in the list is 13 years old. Predicting a price after 20 years would be an extrapolation.

25. Burgers.

a) The scatterplot of calories vs. fat content in fast food hamburgers is at the right. The relationship appears linear, so a linear model is appropriate.

Fat and Calories of Fast Food Burgers

```
Dependent variable is:   Calories
No Selector
R squared = 92.3%     R squared (adjusted) = 90.7%
s =  27.33  with  7 - 2 = 5  degrees of freedom
```

Source	Sum of Squares	df	Mean Square	F-ratio
Regression	44664.3	1	44664.3	59.8
Residual	3735.73	5	747.146	

Variable	Coefficient	s.e. of Coeff	t-ratio	prob
Constant	210.954	50.10	4.21	0.0084
Fat	11.0555	1.430	7.73	0.0006

b) From the computer regression output, $R^2 = 92.3\%$.
92.3% of the variability in the number of calories can be explained by the variability in the number of grams of fat in a fast food burger.

c) From the computer regression output, the regression equation that predicts the number of calories in a fast food burger from its fat content is: $\widehat{Calories} = 210.954 + 11.0555(Fat)$

d) The residuals plot at the right shows no pattern. The linear model appears to be appropriate.

e) The model predicts that a fat free burger would have 210.954 calories. Since there are no data values close to 0, this is an extrapolation outside the data and isn't of much use.

f) For each additional gram of fat in a burger, the model predicts an increase of 11.056 calories.

g) $Cal\hat{o}ries = 210.954 + 11.056(Fat) = 210.954 + 11.0555(28) = 520.508$

The model predicts a burger with 28 grams of fat will have 520.508 calories. If the residual is +33, the actual number of calories is 520.508 + 33 ≈ 553.5 calories.

27. A second helping of burgers.

a) The model from Exercise 25 was for predicting number of calories from number of grams of fat. In order to predict grams of fat from the number of calories, a new linear model needs to be generated.

b) The scatterplot at the right shows the relationship between number fat grams and number of calories in a set of fast food burgers. The association is strong, positive, and linear. Burgers with higher numbers of calories typically have higher fat contents. The relationship is straight enough to apply a linear model.

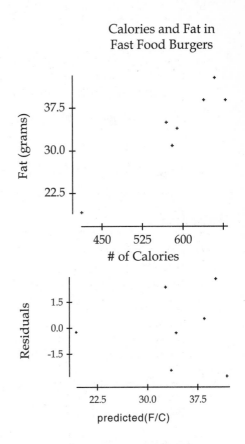

Calories and Fat in Fast Food Burgers

```
Dependent variable is:    Fat
No Selector
R squared = 92.3%     R squared (adjusted) = 90.7%
s =  2.375  with  7 - 2 = 5  degrees of freedom
```

Source	Sum of Squares	df	Mean Square	F-ratio
Regression	337.223	1	337.223	59.8
Residual	28.2054	5	5.64109	

Variable	Coefficient	s.e. of Coeff	t-ratio	prob
Constant	-14.9622	6.433	-2.33	0.0675
Calories	0.083471	0.0108	7.73	0.0006

The linear model for predicting fat from calories is: $F\hat{a}t = -14.9622 + 0.083471(Calories)$ The model predicts that for every additional 100 calories, the fat content is expected to increase by about 8.3 grams.

The residuals plot shows no pattern, so the model is appropriate. $R^2 = 92.3\%$, so 92.3% of the variability in fat content can be explained by the model.

$F\hat{a}t = -14.9622 + 0.083471(Calories)$

$F\hat{a}t = -14.9622 + 0.083471(600)$

$F\hat{a}t \approx 35.1$

According to the model, a burger with 600 calories is expected to have 35.1 grams of fat.

29. Cell phones.

a) $A\hat{n}l = -5.1 + 0.44(200) = 82.9\,\text{minutes}$ of analog talk time is predicted by the model.

b) Since $r = 0.94$, $R^2 = 0.8836$. In other words, 88.36% of the variability in analog talk time can be explained by variability in digital talk time, and the scatterplot is linear. The model should be fairly accurate.

c) The model predicts that for each additional minute of digital talk time, we can expect 0.44 minutes of additional analog talk time.

d) A battery with a positive residual provides more actual talk time than the linear model would predict.

31. El Niño.

a) The correlation between CO_2 level and mean temperature is $r = \sqrt{R^2} = \sqrt{0.334} = 0.5779$.

b) 33.4% of the variability in mean temperature can be explained by variability in CO_2 level.

c) Since the scatterplot of CO_2 level and mean temperature shows a relationship that is straight enough, use of the linear model is appropriate. The linear regression model that predicts mean temperature from CO_2 level is: $Mean\hat{T}emp = 15.3066 + 0.004(CO_2)$

d) The model predicts that an increase in CO_2 level of 1 ppm is associated with an increase of 0.004 °C in mean temperature.

e) According to the model, the mean temperature is predicted to be 15.3066 °C when there is no CO_2 in the atmosphere. This is an extrapolation outside of the range of data, and isn't very meaningful in context, since there is always CO_2 in the atmosphere. We want to use this model to study the change in CO_2 level and how it relates to the change in temperature.

f) The residuals plot shows no apparent patterns. The linear model appears to be an appropriate one.

g)

$$Mean\hat{T}emp = 15.3066 + 0.004(CO_2)$$

$$Mean\hat{T}emp = 15.3066 + 0.004(364)$$

$$Mean\hat{T}emp = 16.7626$$

According to the model, the temperature is predicted to be 16.7626 °C when the CO_2 level is 364 ppm.

33. Body fat.

a) The scatterplot of % body fat and weight of 20 male subjects, at the right, shows a strong, positive, linear association. Generally, as a subject's weight increases, so does % body fat. The association is straight enough to justify the use of the linear model.

The linear model that predicts % body fat from weight is: $\%F\hat{a}t = -27.3763 + 0.249874(Weight)$

Weight and Body Fat

b) The residuals plot, at the right, shows no apparent pattern. The linear model is appropriate.

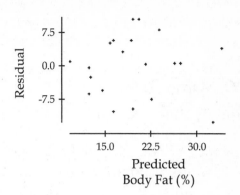

c) According to the model, for each additional pound of weight, body fat is expected to increase by about 0.25%.

d) Only 48.5% of the variability in % body fat can be explained by the model. The model is not expected to make predictions that are accurate.

e)

$$\%F\hat{a}t = -27.3763 + 0.249874(Weight)$$
$$\%F\hat{a}t = -27.3763 + 0.249874(190)$$
$$\%F\hat{a}t \approx 20.09976$$

According to the model, the predicted body fat for a 190-pound man is 20.09976%.
The residual is $21 - 20.09976 \approx 0.9\%$.

35. Hard water.

a) There is a fairly strong, negative, linear relationship between calcium concentration (in ppm) in the water and mortality rate (in deaths per 100,000). Towns with higher calcium concentrations tended to have lower mortality rates.

b) The linear regression model that predicts mortality rate from calcium concentration is
$Mort\hat{a}lity = 1676 - 3.23(Calcium)$.

c) The model predicts a decrease of 3.23 deaths per 100,000 for each additional ppm of calcium in the water. For towns with no calcium in the water, the model predicts a mortality rate of 1676 deaths per 100,000 people.

d) Exeter had 348.6 fewer deaths per 100,000 people than the model predicts.

e)

$$Mort\hat{a}lity = 1676 - 3.23(Calcium)$$
$$Mort\hat{a}lity = 1676 - 3.23(100)$$
$$Mort\hat{a}lity = 1353$$

The town of Derby is predicted to have a mortality rate of 1353 deaths per 100,000 people.

f) 43% of the variability in mortality rate can be explained by variability in calcium concentration.

37. Internet.

The association between the year and the number of Internet journals is very strong, positive, but curved. Later years generally have higher numbers of journals, but the increase per year is not constant. The "straight enough" condition is not satisfied, so the linear model is not appropriate.

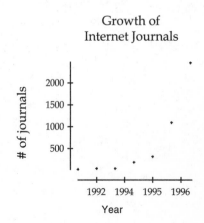

Growth of Internet Journals

Chapter 9 – Regression Wisdom

1. **Marriage age.**

   a) The trend in age at first marriage for American women is very strong over the entire time period recorded on the graph, but the direction and form are different for different time periods. The trend appears to be somewhat linear, and consistent at around 22 years, up until about 1940, when the age seemed to drop dramatically, to under 21. From 1940 to about 1970, the trend appears non-linear and slightly positive. From 1975 to the present, the trend again appears linear and positive. The marriage age rose rapidly during this time period.

   b) The association between age at first marriage for American women and year is strong over the entire time period recorded on the graph, but some time periods have stronger trends than others.

   c) The correlation, or the measure of the degree of linear association is not high for this trend. The graph, as a whole, is non-linear. However, certain time periods, like 1975 to present, have a high correlation.

   d) Overall, the linear model is not appropriate. The scatterplot is not "straight enough" to satisfy the condition. You could fit a linear model to the time period from 1975 to 1995, but this seems unnecessary. The ages for each year are reported, and, given the fluctuations in the past, extrapolation seems risky.

3. **Marriage age revisited.**

   a) Modeling decisions may vary, but the important idea is using a subset of the data that allows us to make an accurate prediction for the year in which we are interested. We might model a subset to predict the marriage age in 2005, and model another subset to predict the marriage age in 1911.

   In order to predict the average marriage age of American women in 2005, use the data points from the most recent trend only. The data points from 1975 – 1995 look straight enough to apply the linear regression model.

Regression output from a computer program is given below, as well as a residuals plot.

```
Dependent variable is:    Age
No Selector
R squared = 97.6%    R squared (adjusted) = 96.9%
s = 0.2470  with  5 - 2 = 3  degrees of freedom

Source        Sum of Squares   df   Mean Square   F-ratio
Regression    7.56900          1      7.56900      124
Residual      0.183000         3      0.061000

Variable    Coefficient   s.e. of Coeff   t-ratio   prob
Constant    -322.430      31.01           -10.4     0.0019
Year        0.174000      0.0156           11.1     0.0016
```

The linear model used to predict average female marriage age from year is:
$A\hat{g}e = -322.430 + 0.174(Year)$. The residuals plot shows no apparent pattern (although the number of points used is small), and the value of R^2 is high. 97.6% of the variability in average female age at first marriage is determined by variability in the year. The model predicts that each year that passes is associated with an increase of 0.174 years in the average female age at first marriage.

$$A\hat{g}e = -322.430 + 0.174(Year)$$
$$A\hat{g}e = -322.430 + 0.174(2005)$$
$$A\hat{g}e \approx 26.44$$

According to the model, the average age at first marriage for women in 2005 will be 26.44 years old. Care should be taken with this prediction, however. It represents an extrapolation of 10 years beyond the highest year.

b) This prediction is for a year that is 10 years higher than the highest year for which we have an average female marriage age. Don't place too much faith in this extrapolation.

c) An extrapolation of more than 50 years into the future would be absurd. There is no reason to believe the trend would continue. In fact, given the situation, it is very unlikely that the pattern would continue in this fashion. The model given in part a) predicts that the marriage age will be 34.27 years. Realistically, that seems quite high.

5. Good model?

a) The student's reasoning is not correct. A scattered residuals plot, not high R^2, is the indicator of an appropriate model. Once the model is deemed appropriate, R^2 is used as a measure of the strength of the model.

b) The model may not allow the student to make accurate predictions. The data may be curved, in which case the linear model would not fit well.

7. Reading.

a) The principal's description of a strong, positive trend is misleading. First of all, "trend" implies a change over time. These data were gathered during one year, at different grade levels. To observe a trend, one class's reading scores would have to be followed through several years. Second, the strong, positive relationship only indicates the yearly improvement that would be expected, as children get older. For example, the 4th graders are reading at approximately a 4th grade level, on average. This means that the school's students are progressing adequately in their reading, not extraordinarily. Finally, the use of average reading scores instead of individual scores increases the strength of the association.

b) The plot appears very straight. The correlation between grade and reading level is very high, probably between 0.9 and 1.0.

c) If the principal had made a scatterplot of all students' scores, the correlation would have likely been lower. Averaging reduced the scatter, since each grade level has only one point instead of many, which inflates the correlation.

d) If a student is reading at grade level, then that student's reading score should equal his or her grade level. The slope of that relationship is 1. That would be "acceptable", according to the measurement scale of reading level. Any slope greater than 1 would indicate above grade level reading scores, which would certainly be acceptable as well. A slope less than 1 would indicate below grade level average scores, which would be unacceptable.

9. Heating.

a) The model predicts a decrease in $2.13 in heating cost for an increase in temperature of 1° Fahrenheit. Generally, warmer months are associated with lower heating costs.

b) When the temperature is 0° Fahrenheit, the model predicts a monthly heating cost of $133.

c) When the temperature is around 32° Fahrenheit, the predictions are generally too high. The residuals are negative, indicating that the actual values are lower than the predicted values.

d)

$\hat{C} = 133 - 2.13(temp)$

$\hat{C} = 133 - 2.13(10)$

$\hat{C} = \$111.70$

According to the model, the heating cost in a month with average daily temperature 10° Fahrenheit is expected to be $111.70.

e) The residual for a 10° day is approximately –$6, meaning that the actual cost was $6 less than predicted, or $111.70 – $6 = $105.70.

f) The model is not appropriate. The residuals plot shows a definite curved pattern. The association between monthly heating cost and average daily temperature is not linear.

g) A change of scale from Fahrenheit to Celsius would not affect the relationship. Associations between quantitative variables are the same, no matter what the units.

11. Unusual points.

a) 1) The point is an outlier, because it is not in the cloud of data points.
2) The point is not influential. It has the *potential* to be influential, because its position far from the mean of the explanatory variable gives it high leverage. However, the point is not *exerting* much influence, because it reinforces the association.
3) If the point were removed, the correlation would become weaker. The point heavily reinforces the positive association. Removing it would weaken the association.
4) The slope would remain roughly the same, since the point is not influential.

b) 1) The point is an outlier. It departs from the overall pattern.
2) The point is influential. The point alone gives the scatterplot the appearance of an overall negative direction, when the points are actually fairly scattered.
3) If the point were removed, the correlation would become weaker. Without the point, there would be very little evidence of linear association.
4) The slope would increase, from a negative slope to a slope near 0. Without the point, the slope of the regression line would be nearly flat.

c) 1) Since it departs from the overall pattern, the point is an outlier.
2) The point is somewhat influential. It is well away from the mean of the explanatory variable, and has enough leverage to change the slope of the regression line, but only slightly.
3) If the point were removed, the correlation would become stronger. Without the point, the positive association would be reinforced.
4) The slope would increase slightly, becoming steeper after the removal of the point. The regression line would follow the general cloud of points more closely.

d) 1) The point is an outlier. It departs from the overall pattern.
2) The point is not influential. It is very close to the mean of the explanatory variable, and the regression line is anchored at the point $(\bar{x}, \bar{y})$, and would only pivot if it were possible to minimize the sum of the squared residuals. No amount of pivoting will reduce the residual for the stray point, so the slope would not change.
3) If the point were removed, the correlation would become slightly stronger, decreasing to become more negative. The point detracts from the overall pattern, and its removal would reinforce the association.
4) The slope would remain roughly the same. Since the point is not influential, its removal would not affect the slope.

13. The extra point.

(1) Point e is very influential. Its addition will give the appearance of a strong, negative correlation like $r = -0.90$.

(2) Point d is influential (but not as influential as point e). Its addition will give the appearance of a weaker, negative correlation like $r = -0.40$.

(3) Point c is directly below the middle of the group of points. Its position is directly below the mean of the explanatory variable. It has no influence. Its addition will leave the correlation the same, $r = 0.00$.

(4) Point b is almost in the center of the group of points, but not quite. Its addition will give the appearance of a very slight positive correlation like $r = 0.05$.

(5) Point a is very influential. Its addition will give the appearance of a strong, positive correlation like $r = 0.75$.

15. Gestation.

a) The association would be stronger if humans were removed. The point on the scatterplot representing human gestation and life expectancy is an outlier from the overall pattern and detracts from the association. Humans also represent an influential point. Removing the humans would cause the slope of the linear regression model to increase, following the pattern of the non-human animals much more closely.

b) The study could be restricted to non-human animals. This appears justifiable, since one could point to a number of environmental factors that could influence human life expectancy and gestation period, making them incomparable to those of non-human animals.

c) The correlation is moderately strong. The model explains 72.2% of the variability in gestation period of non-human animals.

d) For every year increase in life expectancy, the model predicts an increase of approximately 15.5 days in gestation period.

e)

$$\widehat{Gest} = -39.5172 + 15.4980(LifEx)$$
$$\widehat{Gest} = -39.5172 + 15.4980(20)$$
$$\widehat{Gest} \approx 270.4428$$

According to the linear model, monkeys with a life expectancy of 20 years are expected to have gestation periods of about 270.5 days. Care should be taken when assessing the accuracy of this prediction. First of all, the residuals plot has not been examined, so the appropriateness of the model is questionable. Second, it is unknown whether or not monkeys were included in the original 17 non-human species studied. Since monkeys and humans are both primates, the monkeys may depart from the overall pattern as well.

17. Law enforcement.

A scatterplot (with regression line) of assaults (per 1000) and kills/injuries (per 1000) for law enforcement officers authorized to carry firearms is at the right. There is one influential point, the National Park Service (highlighted).

The model we have is highly influenced by the National Park Service, and is of no use as a predictor of kills/injury rates from assault rates.

Below this scatterplot is another, with the influential point removed. Notice how removing the National Park Service changes the slope of the regression line.

There doesn't seem to be a good reason to remove the National Park Service. Although the National Park Service seems like a unique branch of law enforcement, so do all of the others. Each type of officer faces different situations on the job than the others. This is no reason to exclude a particular branch.

Even if we could somehow justify removing the National Park Service, the slope of the regression line is nearly 0, indicating a model with extremely weak predicting ability. In essence, the regression line always predicts the mean killed-injured rate.

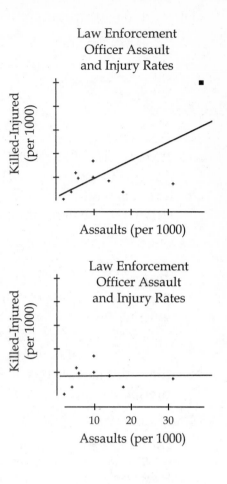

19. Illegitimate births.

The analysis that follows is one of several good models that may be used to predict the percentage of illegitimate births. The important feature to recognize is that these data consist of two distinct trends. Your modeling decisions may vary slightly from these, but that is fine as long as those decisions are justified.

A scatterplot (at the right) of year vs. percent of unmarried births shows two distinct trends. From 1980 to 1994, there is a strong, positive linear association between year and percent of unmarried births. For the years 1995 to 1998, there is also a strong, positive, linear association, but the percent of unmarried births increases much more slowly from year to year, almost to the point of being flat. Two linear models will fit the relationship well.

Model I (1980 – 1994)

Dependent variable is: **Births**
No Selector
R squared = 99.0%   R squared (adjusted) = 98.8%
s = 0.5645 with 7 - 2 = 5 degrees of freedom

Source	Sum of Squares	df	Mean Square	F-ratio
Regression	160.621	1	160.621	504
Residual	1.59314	5	0.318628	

Variable	Coefficient	s.e. of Coeff	t-ratio	prob
Constant	-2021.41	91.25	-22.2	≤ 0.0001
Year	1.02991	0.0459	22.5	≤ 0.0001

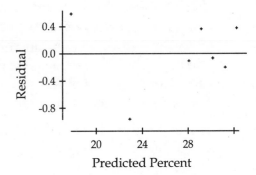

$\hat{\%} = -2021.41 + 1.0299(year)$ is a good model for the years 1980 – 1994. A scatterplot of the relationship, with regression line, is shown above and to the right. $R^2 = 99\%$, so the model explains 99% of the variability in percent of unmarried births. The residuals plot (at the right) is scattered, indicating an appropriate model.

Model II (1995 – 1998)

Dependent variable is: **Births**
No Selector
R squared = 85.3%   R squared (adjusted) = 77.9%
s = 0.1183 with 4 - 2 = 2 degrees of freedom

Source	Sum of Squares	df	Mean Square	F-ratio
Regression	0.162000	1	0.162000	11.6
Residual	0.028000	2	0.014000	

Variable	Coefficient	s.e. of Coeff	t-ratio	prob
Constant	-326.920	105.6	-3.09	0.0905
Year	0.180000	0.0529	3.40	0.0766

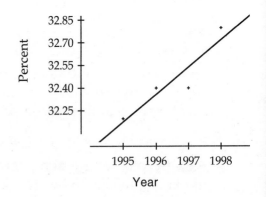

$\hat{\%} = -326.92 + 0.18(year)$ is a good model for the years 1995 – 1998. Although not as accurate as the first model, $R^2 = 85.3\%$, which means that the model accounts for 85.3% of the variability in percent of unmarried births. The residuals plot is scattered, indicating an appropriate model. Great care should be taken in using this model for predictions, since it was developed from only four data points. The slope of the regression line indicates that for each year that passes, the model predicts an increase of only 0.18% unmarried births. The rate may have actually leveled out.

21. Inflation.

a) The trend in Consumer Price Index is strong, non-linear, and positive. Generally, CPI has increased over the years, but the rate of increase has become much greater since approximately 1972. Other characteristics include fluctuations in CPI in the years prior to 1950.

b) In order to effectively predict the CPI in 2010, use only the most recent trend. The trend since 1972 is straight enough to apply the linear model. Prior to 1972, the trend is radically different from that of recent years, and is of no use in predicting CPI for 2010.

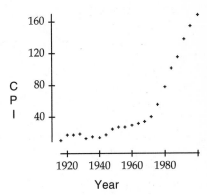

Consumer Price Index

Dependent variable is: **CPI2**
No Selector
R squared = 99.6%   R squared (adjusted) = 99.5%
s = 3.241 with 8 - 2 = 6 degrees of freedom

Source	Sum of Squares	df	Mean Square	F-ratio
Regression	14908.5	1	14908.5	1420
Residual	63.0062	6	10.5010	

Variable	Coefficient	s.e. of Coeff	t-ratio	prob
Constant	-9247.62	248.3	-37.2	≤ 0.0001
Year2	4.71012	0.1250	37.7	≤ 0.0001

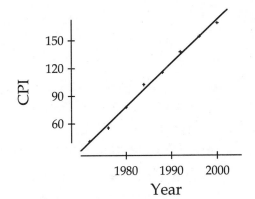

Consumer Price Index

The linear model that predicts CPI from year is
$\hat{CPI} = -9247.62 + 4.71012(year)$. $R^2 = 99.6\%$, meaning that the model predicts 99.6% of the variability in CPI. The residuals plot shows no pattern, so the linear model is appropriate. According to the model, the CPI is expected to increase by $4.70 each year, for 1972 − 2000.

$\hat{CPI} = -9247.62 + 4.71012(year)$

$\hat{CPI} = -9247.62 + 4.71012(2010)$

$\hat{CPI} \approx 219.72$

In 2010, CPI is predicted to be $219.72

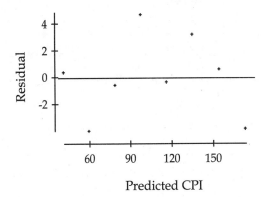

c) In the scatterplot of year versus log(CPI), differences in CPI are accentuated. We can clearly see the impact on CPI of historical events such as the stock market crash of 1929, the Great Depression in the years following, and World War II.

Consumer Price

Chapter 10 – Re-expressing Data:
It's Easier Than You Think

1. **Models.**

   a) $\hat{y} = 1.2 + 0.8x$
   $\hat{y} = 1.2 + 0.8(2)$
   $\hat{y} = 2.8$

   b) $\ln \hat{y} = 1.2 + 0.8x$
   $\ln \hat{y} = 1.2 + 0.8(2)$
   $\ln \hat{y} = 2.8$
   $\hat{y} = e^{2.8} = 16.44$

   c) $\sqrt{\hat{y}} = 1.2 + 0.8x$
   $\sqrt{\hat{y}} = 1.2 + 0.8(2)$
   $\sqrt{\hat{y}} = 2.8$
   $\hat{y} = 2.8^2 = 7.84$

   d) $\dfrac{1}{\hat{y}} = 1.2 + 0.8x$
   $\dfrac{1}{\hat{y}} = 1.2 + 0.8(2)$
   $\dfrac{1}{\hat{y}} = 2.8$
   $\hat{y} = \dfrac{1}{2.8} = 0.36$

   e) $\hat{y} = 1.2(2)^{0.8}$
   $\hat{y} = 2.09$

3. **Gas mileage.**

   a) The association between weight and gas mileage of cars is fairly linear, strong, and negative. Heavier cars tend to have lower gas mileage.

   b) For each additional thousand pounds of weight, the linear model predicts a decrease of 7.652 miles per gallon in gas mileage.

   c) The linear model is not appropriate. There is a curved pattern in the residuals plot. The model tends to underestimate gas mileage for cars with relatively low and high gas mileages, and overestimates the gas mileage of cars with average gas mileage.

   d) The residuals plot for the re-expressed relationship is much more scattered. This is an indication of an appropriate model.

   e) The linear model that predicts the number of gallons per 100 miles in gas mileage from the weight of a car is: $\widehat{Gal} / 100 = 0.625 + 1.178(Weight)$.

   f) For each additional 1000 pounds of weight, the model predicts that the car will require an additional 1.178 gallons to drive 100 miles.

   g)

   $\widehat{Gal} / 100 = 0.625 + 1.178(Weight)$
   $\widehat{Gal} / 100 = 0.625 + 1.178(3.5)$
   $\widehat{Gal} / 100 = 4.748$

   According to the model, a car that weighs 3500 pounds (3.5 thousand pounds) is expected to require approximately 4.748 gallons to drive 100 miles, or 0.04748 gallons per mile.

   This is $\dfrac{1}{0.04748} \approx 21.06$ miles per gallon.

5. Brakes.

a) The association between speed and stopping distance is strong, positive, and appears straight. Higher speeds are generally associated with greater stopping distances. The linear regression model, with equation $\widehat{Distance} = -65.9 + 5.98(Speed)$, has $R^2 = 96.9\%$, meaning that the model explains 96.9% of the variability in stopping distance. However, the residuals plot has a curved pattern. The linear model is not appropriate. A model using re-expressed variables should be used.

Stopping Distance and Speed

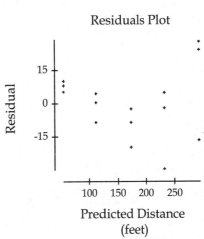

Residuals Plot

b) Stopping distances appear to be relatively higher for higher speeds. This increase in the rate of change might be able to be straightened by taking the square root of the response variable, stopping distance. The scatterplot of Speed versus $\sqrt{Distance}$ seems like it might be a bit straighter.

Square Root Re-expression

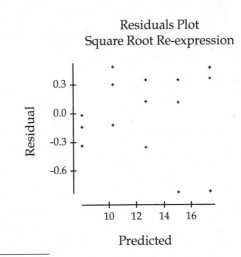

Residuals Plot Square Root Re-expression

c) The model for the re-expressed data is $\sqrt{\widehat{Distance}} = 3.303 + 0.235(Speed)$. The residuals plot shows no pattern, and $R^2 = 98.4\%$, so 98.4% of the variability in the square root of the stopping distance can be explained by the model.

d)

$$\sqrt{Dis\hat{t}ance} = 3.303 + 0.235(Speed)$$

$$\sqrt{Dis\hat{t}ance} = 3.303 + 0.235(55)$$

$$\sqrt{Dis\hat{t}ance} = 16.228$$

$$Dis\hat{t}ance = 16.228^2 \approx 263.4$$

According to the model, a car traveling 55 mph is expected to require approximately 263.4 feet to come to a stop.

e)

$$\sqrt{Dis\hat{t}ance} = 3.303 + 0.235(Speed)$$

$$\sqrt{Dis\hat{t}ance} = 3.303 + 0.235(70)$$

$$\sqrt{Dis\hat{t}ance} = 19.753$$

$$Dis\hat{t}ance = 19.753^2 \approx 390.2$$

According to the model, a car traveling 70 mph is expected to require approximately 390.2 feet to come to a stop.

f) The level of confidence in the predictions should be quite high. R^2 is high, and the residuals plot is scattered. The prediction for 70 mph is a bit of an extrapolation, but should still be reasonably close.

7. Baseball salaries.

a) The scatterplot of year versus highest salary is moderately strong, positive and curved. The highest salary has generally increased over the years. Since the scatterplot shows a curved relationship, the linear model is not appropriate. Using the logarithm re-expression on the salaries seems to straighten the scatterplot considerably.

b) $\log(Sal\hat{a}ry) = -113.473 + 0.05734(Year)$ is a good mode for predicting the highest salary from the year. The residuals plot is scattered and $R^2 = 84.9\%$, indicating that 84.9% of the variability in highest salary can be explained by the model.

c) $\log(Sal\hat{a}ry) = -113.473 + 0.05734(Year) = -113.473 + 0.05734(2005) = 1.4937$

$Sal\hat{a}ry = 10^{1.4937} \approx 31.2$

According to the model, the highest salary in baseball is expected to be approximately 31.2 million dollars. (Be careful with decimal places here. When the slope is small and the value of the explanatory variable is large, like a year, too much rounding of the slope can be detrimental to the accuracy of predictions. When using logs, this is an even bigger problem!)

9. Planet distances and order.

a) The association between planetary position and distance from the sun is strong, positive, and curved (below left). A good re-expression of the data is position versus Log(Distance). The scatterplot with regression line (below center) shows the straightened association. The equation of the model is $\log(Dis\hat{t}ance) = 1.245 + 0.271(Position)$. The residuals plot (below right) may have some pattern, but after trying several re-expressions, this is the best that can be done. $R^2 = 98.2\%$, so the model explains 98.2% of the variability in the log of the planets distance from the sun.

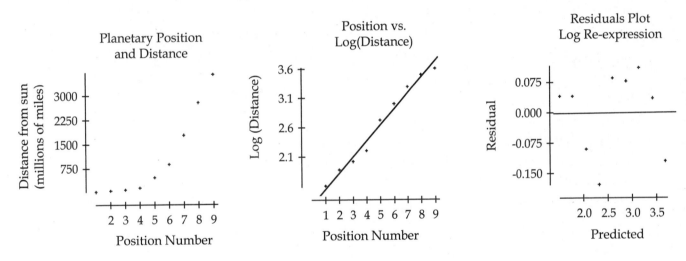

b) At first glance, this model appears to provide little evidence to support the contention of the astronomers. Pluto appears to fit the pattern, although Pluto's distance from the sun is a bit less than expected. A model generated without Pluto does not have a dramatically improved residuals plot, does not have a significantly higher R^2, nor a different slope. Pluto does not appear to be influential.

But don't forget that a logarithmic scale is being used for the vertical axis. The higher up the vertical axis you go, the greater the effect of a small change.

$\log(Dis\hat{t}ance) = 1.24521 + 0.270922(Position)$

$\log(Dis\hat{t}ance) = 1.24521 + 0.270922(9)$

$\log(Dis\hat{t}ance) = 3.683508$

$Dis\hat{t}ance = 10^{3.683508} \approx 4825$

According to the model, the 9th planet in the solar system is predicted to be approximately 4825 million miles away from the sun. Pluto is actually 3666 million miles away.

Pluto doesn't fit the pattern for position and distance in the solar system. In fact, the model made with Pluto included isn't a good one, because Pluto influences those predictions. The model without Pluto, $\log(Dis\hat{t}ance) = 1.20259 + 0.283706(Position)$, works much better. It has a high R^2, and scattered residuals plot. This new model predicts that the 9th planet should be a whopping 5701 million miles away from the sun! There is evidence that the astronomers are correct. Pluto doesn't behave like planet in its relation to position and distance.

11. Quaoar, planets, part 4.

a)

$$\log(Dis\hat{t}ance) = 1.20259 + 0.283706(Position)$$

$$\log(Dis\hat{t}ance) = 1.20259 + 0.283706(9)$$

$$\log(Dis\hat{t}ance) = 3.755944$$

$$Dis\hat{t}ance = 10^{3.755944} \approx 5701$$

Using the model predicting distance from the sun from position, the 9th planet is expected to be about 5701 million miles from the sun. At 4000 million miles, Quaoar is a better fit than Pluto, but still not a good fit.

b) With Quaoar as the 9th planet, the model that relates planetary position to distance is $\log(Dis\hat{t}ance) = 1.23679 + 0.273447(Position)$.

$$\log(Dis\hat{t}ance) = 1.23679 + 0.273447(Position)$$

$$\log(Dis\hat{t}ance) = 1.23679 + 0.273447(9)$$

$$\log(Dis\hat{t}ance) = 3.697813$$

$$Dis\hat{t}ance = 10^{3.697813} \approx 4987$$

According to the new model, the 9th planet is predicted to be 4987 million miles from the sun. Quaoar, at approximately 4000 million miles, is a better fit than Pluto, but doesn't appear to be an additional planet.

13. Logs (not logarithms).

a) The association between the diameter of a log and the number of board feet of lumber is strong, positive, and curved. As the diameter of the log increases, so does the number of board feet of lumber contained in the log.

The model used to generate the table used by the log buyers is based upon a square root re-expression. The values in the table correspond exactly to the model

$$\sqrt{Boar\hat{d}Feet} = -4 + Diameter.$$

b)

$$\sqrt{Boar\hat{d}Feet} = -4 + Diameter$$

$$\sqrt{Boar\hat{d}Feet} = -4 + (10)$$

$$\sqrt{Boar\hat{d}Feet} = 6$$

$$Boar\hat{d}Feet = 36$$

According to the model, a log 10" in diameter is expected to contain 36 board feet of lumber.

c)

$$\sqrt{Boar\hat{d}Feet} = -4 + Diameter$$

$$\sqrt{Boar\hat{d}Feet} = -4 + (36)$$

$$\sqrt{Boar\hat{d}Feet} = 32$$

$$Boar\hat{d}Feet = 1024$$

According to the model, a log 36" in diameter is expected to contain 1024 board feet of lumber.

Normally, we would be cautious of this prediction, because it is an extrapolation beyond the given data, but since this is a prediction made from an exact model based on the volume of the log, the prediction will be accurate.

15. Life expectancy.

The association between year and life expectancy is strong, curved and positive. As the years passed, life expectancy for white males has increased. The bend (in mathematical language, the change in curvature) in the scatterplot makes it impossible to straighten the association by re-expression.

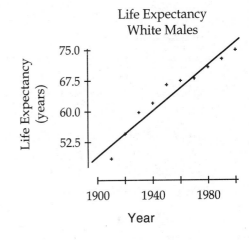

If all of the data points are to be used, the linear model, $Life\hat{E}xp = -454.938 + 0.265697(Year)$ is probably the best model we can make. It doesn't fit the data well, but accounts for 93.6% of the variability in life expectancy.

In order to make better future predictions, we might use only a subset of the data points. The scatterplot is very linear from 1970 to 2000, and these recent years are likely to be more indicative of the current trend in life expectancy. Using only these four data points, we might be able to make more accurate future predictions.

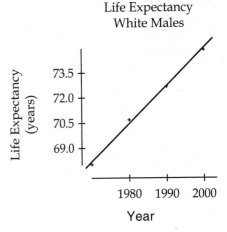

The model $Life\hat{E}xp = -379.020 + 0.227(Year)$ explains 99.6% of the variability in life expectancy, and has a scattered residuals plot. Great care should be taken in using this model for predictions, since it is based upon only 4 points, and, as always, prediction into the future is risky. This is especially true since the response variable is life expectancy, which cannot be expected to increase at the same rate indefinitely.

17. Slower is cheaper.

Speed and Mileage

The association between speed and mileage is strong and curved. For speeds under 50 miles per hour, higher speeds are associated with higher mileage. For speeds in excess of 50 mph, higher speeds are associated with lower mileage. For this reason, the scatterplot cannot be straightened by the methods used in this chapter.

The most logical modeling decision at this point is to divide the curve up into two linear models. The association seems fairly linear from 35−50 mph, and also fairly linear from 55−75 mph.

For lower speeds, the model is $Mil\hat{e}age = 18.040 + 0.232(Speed)$. The model has some pattern in the residuals plot, but no re-expression seems to change it. The model explains 96.7% of the variability in mileage. For low speeds, the model predicts an increase of 0.232 miles per gallon for each additional mile per hour of speed.

For high speed, the model is $Mil\hat{e}age = 46.800 − 0.320(Speed)$. The residuals plot is scattered and the model explains 99.1% of the variability in mileage. For higher speeds, the model predicts a decrease in mileage of 0.320 miles per gallon for each additional mile per hour of speed

Mileage at Lower Speeds

Residuals Plot 35−50 mph

Mileage at Higher Speeds

Residuals Plot 55−75 mph

19. Years to live.

a) The association between the age and estimated
additional years of life for black males is strong, curved,
and negative. Older men generally have fewer estimated
years of life remaining. The square root re-expression of
the data, $\sqrt{Year\hat{s}Left} = 8.39253 - 0.069553(Age)$,
straightens the data considerably, but has an extremely
patterned residuals plot. The model is not a
mathematically appropriate model, but fits so closely that
it should be fine for predictions within the range of data.
The model explains 99.8% of the variability in the
estimated number of additional years.

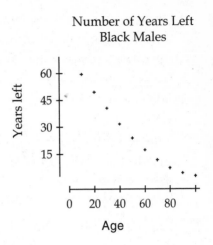

Number of Years Left
Black Males

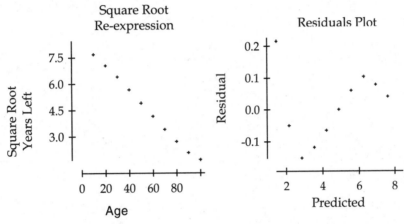

b)

$$\sqrt{Year\hat{s}Left} = 8.39253 - 0.069553(Age)$$

$$\sqrt{Year\hat{s}Left} = 8.39253 - 0.069553(18)$$

$$\sqrt{Year\hat{s}Left} = 7.140936$$

$$Year\hat{s}Left = 7.140936^2 \approx 50.99$$

According to the model, an 18-year-old
black male is expected to live and additional
50.99 years, for a total age of 68.99 years.

21. Internet.

a) The association between the year and the number of electronic journals is strong, positive, and curved. The number of electronic journals rapidly increased in the last several years. Clearly, the linear model would not be appropriate for modeling these data.

A good model is the log re-expression $\text{Log}(Jou\hat{r}nals) = -686.763 + 0.345547(Year)$. The residuals plot is scattered, and the value of R^2 indicates that the model explains 95.9% of the variability in the number of journals.

Internet Journals

Log Re-expression

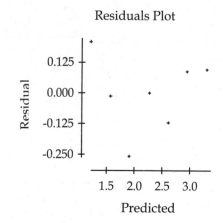

Residuals Plot

b)

$\text{Log}(Jou\hat{r}nals) = -686.763 + 0.345547(Year)$

$\text{Log}(Jou\hat{r}nals) = -686.763 + 0.345547(2000)$

$\text{Log}(Jou\hat{r}nals) = 4.334$

$Jou\hat{r}nals = 10^{4.334} \approx 21,429$

According to the model, there are expected to be approximately 21,429 electronic journals in the year 2000.

c) The estimate of 21,429 journals in 2000 seems very high, and seems unlikely to be accurate. The year 2000 is an extrapolation 3 years beyond the most recent year included in the data set. The model is increasing exponentially, but it is quite unlikely that the number of journals could keep increasing exponentially.

Review of Part II – Exploring Relationships Between Variables

1. **College.**

% over 50: $r = 0.69$	The only moderate, positive correlation in the list.
% under 20: $r = -0.71$	Moderate, negative correlation (-0.98 is too strong)
% Gr. on time: $r = -0.51$	Moderate, negative correlation (not as strong as %under 20)
% Full-time Fac.: $r = 0.09$	No correlation.

3. **Wines.**

   a) There does not appear to be an association between ages of vineyards and the cost of wines. $r = \sqrt{R^2} = \sqrt{0.027} = 0.164$, indicating a very weak association, at best. The model only explains 2.7% of the variability in case price. Furthermore, the regression equation appears to be influenced by two outliers, wines from vineyards over 30 years old, with relatively high case prices.

   b) This analysis tells us nothing about wines worldwide. There is no reason to believe that the results for the Finger Lakes region are representative of the wines of the world.

   c) The linear equation used to predict case price from age of the vineyard is:
   $$Case\hat{P}rice = 92.765 + 0.567284(Years)$$

   d) This model is not useful because only 2.7% of the variability in case price is accounted for by the ages of the vineyards. Furthermore, the slope of the regression line seems influenced by the presence of two outliers, wines from vineyards over 30 years old, with relatively high case prices.

5. **More twins?**

   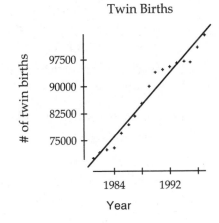

   a) The association between year and the number of twin births is strong, positive, and appears non-linear. Generally, the number of twin births has increased over the years. The linear model that predicts the number of twin births from the year is:
   $$Tw\hat{i}ns = -4316980 + 2214.19(Year)$$

   b) For each year that passes, the model predicts that the number of twins born will increase by approximately 2214 twin births.

   c)
   $$Tw\hat{i}ns = -4,316,980 + 2214.19(Year)$$
   $$Tw\hat{i}ns = -4,316,980 + 2214.19(2002)$$
   $$Tw\hat{i}ns \approx 115,828.4$$

   According to the model, there are expected to be 115,828.4 twin births in the US in 2002. However, the scatterplot appears non-linear, and there is no reason to believe the number of twin births will keep increasing at the same rate for 5 years beyond the last recorded year. Faith in this prediction is very low.

d) The residuals plot shows a definite curved pattern. The association is not linear, so the linear model is not appropriate.

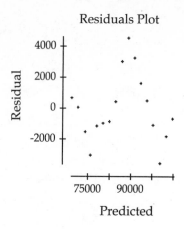

Residuals Plot

7. Acid rain.

a) $r = \sqrt{R^2} = \sqrt{0.27} = -0.5196$. The association between pH and BCI appears negative in the scatterplot, so use the negative value of the square root.

b) The association between pH and BCI is negative, moderate, and linear. Generally, higher pH is associated with lower BCI. Additionally, BCI appears more variable for higher values of pH.

c) In a stream with average pH, the BCI would be expected to be average, as well.

d) In a stream where the pH is 3 standard deviations above average, the BCI is expected to be 1.56 standard deviations below the mean level of BCI. (r (3) = $-0.5196(3) = -1.56$)

9. A Manatee model.

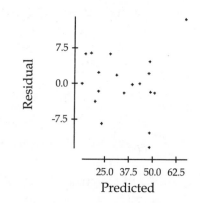

Residuals Plot

a) The association between the number of powerboat registrations and the number of manatees killed is straight enough to try a linear model.

$\widehat{Kills} = -45.8933 + 0.131759(ThousandBoats)$ is the best fitting model. The residuals plot is scattered, so the linear model is appropriate.

b) For every additional 10,000 powerboats registered, the model predicts that an additional 1.32 manatees will be killed.

c) The model predicts that if no powerboats were registered, the number of manatee deaths would be approximately -46. This is an extrapolation beyond the scope of the data, and doesn't have much contextual meaning.

d)

$\widehat{Kills} = -45.8933 + 0.131759(ThousandBoats)$

$\widehat{Kills} = -45.8933 + 0.131759(860)$

$\widehat{Kills} \approx 67.4$

The model predicted 67.4 manatee kills in 2001, when the number of powerboat registrations was 860,000. The actual number of kills was 81. The model under-predicted the number of kills by 13.6.

e) Negative residuals are better for the manatees. A negative residual suggests that the actual number of kills was below the number of kills predicted by the model.

f) Over time, the number of powerboat registrations has increased and the number of manatees killed has increased. The trend may continue, resulting in a greater number of manatee deaths in the future.

11. Traffic.

a)

$$b_1 = r\left(\frac{s_y}{s_x}\right)$$

$$-0.352 = r\left(\frac{9.68}{27.07}\right)$$

$$r = -0.984$$

The correlation between traffic density and speed is $r = -0.984$

b) $R^2 = (-0.984)^2 = 0.969$.
The variation in the traffic density explains 96.9% of the variation in speed.

c)

$$\hat{speed} = 50.55 - 0.352cars$$

$$\hat{speed} = 50.55 - 0.352(50)$$

$$\hat{speed} = 32.95$$

According to the linear model, when traffic density is 50 cars per mile, the average speed of traffic on a moderately large city thoroughfare is expected to be 32.95 miles per hour.

d)

$$\hat{speed} = 50.55 - 0.352cars$$

$$\hat{speed} = 50.55 - 0.352(56)$$

$$\hat{speed} = 30.84$$

According to the linear model, when traffic density is 56 cars per mile, the average speed of traffic on a moderately large city thoroughfare is expected to be 30.84 miles per hour. If traffic is actually moving at 32.5 mph, the residual is $32.5 - 30.84 = 1.66$ miles per hour.

e)

$$\hat{speed} = 50.55 - 0.352cars$$

$$\hat{speed} = 50.55 - 0.352(125)$$

$$\hat{speed} = 6.55$$

According to the linear model, when traffic density is 125 cars per mile, the average speed of traffic on a moderately large city thoroughfare is expected to be 6.55 miles per hour. The point with traffic density 125 cars per minute and average speed 55 miles per hour is considerably higher than the model would predict. If this point were included in the analysis, the slope would increase.

f) The correlation between traffic density and average speed would become weaker. The influential point (125, 55) is a departure from the pattern established by the other data points.

g) The correlation would not change if kilometers were used instead of miles in the calculations. Correlation is a "unitless" measure of the degree of linear association based on z-scores, and is not affected by changes in scale. The correlation would remain the same, $r = -0.984$.

13. Correlations.

a) Weight, with a correlation of –0.903, seems to be most strongly associated with fuel economy, since the correlation has the largest magnitude (distance from zero). However, without looking at a scatterplot, we can't be sure that the relationship is linear. Correlation might not be an appropriate measure of the strength of the association if the association is non-linear.

b) The negative correlation between weight and fuel economy indicates that, generally, cars with higher weights tend to have lower mileages than cars with lower weights. Once again, this is only correct if the association between weight and fuel economy is linear.

c) $R^2 = (-0.903)^2 = 0.815$. The variation in weight accounts for 81.5% of the variation in mileage. Once again, this is only correct if the association between weight and fuel economy is linear.

15. Cars, one more time.

a) The linear model that predicts the horsepower of an engine from the weight of the car is:
$\widehat{Horsepower} = 3.49834 + 34.3144(Weight)$.

b) The weight is measured in thousands of pounds. The slope of the model predicts an increase of about 34.3 horsepower for each additional unit of weight. 34.3 horsepower for each additional thousand pounds makes more sense than 34.3 horsepower for each additional pound.

c) Since the residuals plot shows no pattern, the linear model is appropriate for predicting horsepower from weight.

d)

$\widehat{Horsepower} = 3.49843 + 34.3144(Weight)$

$\widehat{Horsepower} = 3.49843 + 34.3144(2.595)$

$\widehat{Horsepower} \approx 92.544$

According to the model, a car weighing 2595 pounds is expected to have 92.543 horsepower. The actual horsepower of the car is: $92.544 + 22.5 \approx 115.0$ horsepower.

17. Old Faithful.

a) The association between the duration of eruption and the interval between eruptions of Old Faithful is fairly strong, linear, and positive. Long eruptions are generally associated with long intervals between eruptions. There are also two distinct clusters of data, one with many short eruptions followed by short intervals, the other with many long eruptions followed by long intervals, with only a few medium eruptions and intervals in between.

b) The linear model used to predict the interval between eruptions is:
$\widehat{Interval} = 33.9668 + 10.3582(Duration)$.

c) As the duration of the previous eruption increases by one minute, the model predicts an increase of about 10.4 minutes in the interval between eruptions.

d) $R^2 = 77.0\%$, so the model accounts for 77% of the variability in the interval between eruptions. The predictions should be fairly accurate, but not precise. Also, the association appears linear, but we should look at the residuals plot to be sure that the model is appropriate before placing too much faith in any prediction.

e)

$Inte\hat{r}val = 33.9668 + 10.3582(Duration)$

$Inte\hat{r}val = 33.9668 + 10.3582(4)$

$Inte\hat{r}val \approx 75.4$

According to the model, if an eruption lasts 4 minutes, the next eruption is expected to occur in approximately 75.4 minutes.

f) The actual eruption at 79 minutes is 3.6 minutes later than predicted by the model. The residual is 79 − 75.4 = 3.6 minutes. In other words, the model under-predicted the interval.

19. How old is that tree?

a) The correlation between tree diameter and tree age is $r = 0.888$. Although the correlation is moderately high, this does not suggest that the linear model is appropriate. We must look at a scatterplot in order to verify that the relationship is straight enough to try the linear model. After finding the linear model, the residuals plot must be checked. If the residuals plot shows no pattern, the linear model can be deemed appropriate.

b) The association between diameter and age of these trees is fairly strong, somewhat linear, and positive. Trees with larger diameters are generally older.

c) The linear model that predicts age from diameter of trees is: $A\hat{g}e = -0.974424 + 2.20552(Diameter)$. This model explains 78.9% of the variability in age of the trees.

d) The residuals plot shows a curved pattern, so the linear model is not appropriate. Additionally, there are several trees with large residuals.

e) The largest trees are generally above the regression line, indicating a positive residual. The model is likely to underestimate these values.

Tree Age and Diameter

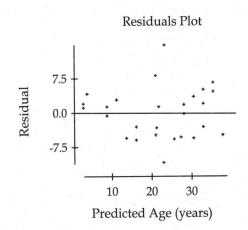

Residuals Plot

21. New homes.

New homes might have areas somewhere between 1000 and 5000 square feet. Doubling the area of a house from 1000 square feet to 2000 square feet would be predicted to increase the price by either 0.008(1000) = $8 thousand, 0.08(1000) = $80 thousand, 0.8(1000) = $800 thousand, or 8(1000) = $8000 thousand. The only reasonable answer is $80 thousand, so the slope must be 0.08.

23. No smoking?

a) The model from Exercise 22 is for predicting the percent of expectant mothers who smoked during their pregnancies from the year, not the year from the percent.

b) The model that predicts the year from the percent of expectant mothers who smoked during pregnancy is: $\hat{Year} = 2009.92 - 1.12968(\%)$. This model predicts that 0% of mothers will smoke during pregnancy in $2009.92 - 1.12968(0) = 2009.92 \approx 2010$.

c) The lowest data point corresponds to 10.4% of expectant mothers smoking during pregnancy in 1998. The prediction for 0% is an extrapolation outside the scope of the data. There is no reason to believe that the model will be accurate at that point.

25. Move south?

There is a strong, roughly linear, negative association between mean January temperature and latitude. U.S. cities with higher latitudes generally have lower mean January temperatures. There are two outliers, cities with higher mean January temperatures than the pattern would suggest.

27. Winter in the city.

a) $R^2 = (-0.848)^2 \approx 0.719$. The variation in latitude explains 71.9% of the variability in average January temperature.

b) The negative correlation indicates that the as latitude increases, the average January temperature generally decreases.

c)

$$b_1 = r\left(\frac{s_y}{s_x}\right)$$

$$b_1 = -0.848\left(\frac{13.49}{5.42}\right)$$

$$b_1 = -2.1106125$$

$$\hat{y} = b_0 + b_1 x$$
$$\bar{y} = b_0 + b_1\bar{x}$$
$$26.44 = b_0 - 2.1106125(39.02)$$
$$b_0 = 108.79610$$

The equation of the linear model for predicting January temperature from latitude is:
$$\hat{JanTemp} = 108.796 - 2.111(Latitude)$$

d) For each additional degree of latitude, the model predicts a decrease of approximately 2.1°F in average January temperature.

e) The model predicts that the mean January temperature will be approximately 108.8°F when the latitude is 0°. This is an extrapolation, and may not be meaningful.

f)

$$Jan\hat{T}emp = 108.796 - 2.111(Latitude)$$

$$Jan\hat{T}emp = 108.796 - 2.111(40) \approx 24.4$$

According to the model, the mean January temperature in Denver is expected to be 24.4°F.

g) In this context, a positive residual means that the actual average temperature in the city was higher than the temperature predicted by the model. In other words, the model underestimated the average January temperature.

29. Jumps.

a) The association between Olympic long jump distances and high jump heights is strong, linear, and positive. Years with longer long jumps tended to have higher high jumps.

b) There is an association between long jump and high jump performance, but it is likely that training and technique have improved over time and affected both jump performances.

c) The correlation would be the same, 0.92. Correlation is a measure of the degree of linear association between two quantitative variables and is unaffected by changes in units.

d) In a year when the high jumper jumped one standard deviation better than the average jump, the long jumper would be predicted to jump $r = 0.92$ standard deviations above the average long jump.

31. French.

a) Most of the students would have similar weights. Regardless of their individual French vocabularies, the correlation would be near 0.

b) There are two possibilities. If the school offers French at all grade levels, then the correlation would be positive and strong. Older students, who typically weigh more, would have higher scores on the test, since they would have learned more French vocabulary. If French is not offered, the correlation between weight and test score would be near 0. Regardless of weight, most students would have horrible scores.

c) The correlation would be near 0. Most of the students would have similar weights and vocabulary test scores. Weight would not be a predictor of score.

d) The correlation would be positive and strong. Older students, who typically weigh more, would have higher test scores, since they would have learned more French vocabulary.

33. Lunchtime.

The association between time spent at the table and number of calories consumed by toddlers is moderate, roughly linear, and negative. Generally, toddlers who spent a longer time at the table consumed fewer calories than toddlers who left the table quickly. The scatterplot between time at the table and calories consumed is straight enough to justify the use of the linear model. The linear model that predicts the time number of calories consumed by a toddler from the time spent at the table is $Cal\hat{o}ries = 560.7 - 3.08(Time)$. For each additional minute spent at the table, the model predicts that the number of calories consumed will be approximately 3.08 fewer. Only 42.1% of the variability in the number of calories consumed can be explained by the variability in time spent at the table. The residuals plot shows no pattern, so the linear model is appropriate, if not terribly useful for prediction.

35. Tobacco and alcohol.

The first concern about these data is that they consist of averages for regions in Great Britain, not individual households. Any conclusions reached can only be about the regions, not the individual households living there. The second concern is the data point for Northern Ireland. This point has high leverage, since it has the highest household tobacco spending and the lowest household alcohol spending. With this point included, there appears to be only a weak positive association between tobacco and alcohol spending. Without the point, the association is much stronger. In Great Britain, with the exception of Northern Ireland, higher levels of household spending on tobacco are associated with higher levels of household spending on tobacco. It is not necessary to make the linear model, since we have the household averages for the regions in Great Britain, and the model wouldn't be useful for predicting in other countries or for individual households in Great Britain.

37. Models.

a)

$$\hat{y} = 2 + 0.8 \ln x$$
$$\hat{y} = 2 + 0.8 \ln(10)$$
$$\hat{y} \approx 3.842$$

b)

$$\log \hat{y} = 5 - 0.23x$$
$$\log \hat{y} = 5 - 0.23(10)$$
$$\log \hat{y} = 2.7$$
$$\hat{y} = 10^{2.7} \approx 501.187$$

c)

$$\frac{1}{\sqrt{\hat{y}}} = 17.1 - 1.66x$$
$$\frac{1}{\sqrt{\hat{y}}} = 17.1 - 1.66(10) = 0.5$$
$$\hat{y} = \frac{1}{0.5^2} = 4$$

39. Vehicle weights.

a)

$$\hat{Wt} = 10.85 + 0.64 scale$$
$$\hat{Wt} = 10.85 + 0.64(31.2)$$
$$\hat{Wt} = 30.818$$

According to the model, a truck with a scale weight of 31,200 pounds is expected to weigh 30,818 pounds.

b) If the actual weight of the truck is 32,120 pounds, the residual is 32,120 – 30,818 = 1302 pounds. The model underestimated the weight.

c)

$$\hat{Wt} = 10.85 + 0.64 scale$$
$$\hat{Wt} = 10.85 + 0.64(35.590)$$
$$\hat{Wt} = 33.6276 \text{ thousand pounds}$$

The predicted weight of the truck is 33,627.6 pounds. If the residual is –2440 pounds, the actual weight of the truck is 33,627.6 – 2440 = 31,187.6 pounds.

d) $R^2 = 93\%$, so the model explains 93% of the variability in weight, but some of the residuals are 1000 pounds or more. If we need to be more accurate than that, then this model will not work well.

e) Negative residuals will be more of a problem. Police would be issuing tickets to trucks whose weights had been overestimated by the model. The U.S. justice system is based upon the principle of innocence until guilt is proven. These truckers would be unfairly ticketed, and that is worse than allowing overweight trucks to pass.

41. Down the drain.

The association between diameter of the drain plug and drain time of this water tank is strong, curved, and negative. Tanks with larger drain plugs have lower drain times. The linear model is not appropriate for the curved association, so several re-expressions of the data were tried. The best one was the reciprocal square root re-expression, resulting in the equation $\dfrac{1}{\sqrt{Drain\hat{T}ime}} = 0.00243 + 0.219(Diameter)$.

The re-expressed data is nearly linear, and although the residuals plot might still indicate some pattern and has one large residual, this is the best of the models examined. The model explains 99.7% of the variability in drain time.

43. Least Squares.

If the 4 x-values are plugged into $\hat{y} = 810 - 3x$, the 4 predicted values are $\hat{y} = 774$, 738, 702 and 666, respectively. The 4 residuals are 26, –58, 38, –6. The squared residuals are 676, 3364, 1444, and 36, respectively. The sum of the squared residuals is 5520. Least squares means that no other line has a sum lower than 5520. In other words, it's the best fit.

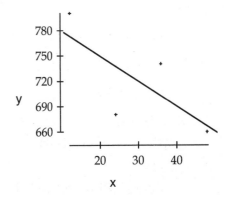

Chapter 11 – Understanding Randomness

1. **Coin toss.**

   A coin flip is random, because the outcome cannot be predicted beforehand.

3. **The lottery.**

   In state lotteries, a machine pops up numbered balls. If the lottery were truly random, the outcome could not be predicted and the outcomes would be equally likely. It is random only if the balls generate numbers in equal frequencies.

5. **Bad simulations**

   a) The outcomes are not equally likely. For example, the probability of getting 5 heads in 9 tosses is not the same as the probability of getting 0 heads, but the simulation assumes they are equally likely.

   b) The even-odd assignment assumes that the player is equally likely to score or miss the shot. In reality, the likelihood of making the shot depends on the player's skill.

   c) Suppose a hand has four aces. This might be represented by 1,1,1,1, and any other number. The likelihood of the first ace in the hand is not the same as for the second or third or fourth. But with this simulation, the likelihood is the same for each.

7. **Wrong conclusion.**

   The conclusion should indicate that the simulation **suggests** that the average length of the line would be 3.2 people. Future results might not match the simulated results exactly.

9. **Cereal.**

   Answers will vary. A component is the simulation of the picture in one box of cereal. One possible way to model this component is to generate random digits 0-9. Let 0 and 1 represent Tiger Woods, 2-4 represent Lance Armstrong, and 5-9 represent Serena Williams. Each run will consist of 5 random digits, and the response variable will be whether or not a complete set of pictures is simulated. Runs in which at least one of each picture is simulated will be a success. The total number of successes divided by the total number of runs will be the simulated probability of ending up with a complete set of pictures. According to the simulation, the probability of getting a complete set of pictures is expected to be about 51.5%.

11. **Multiple choice**

    Answers will vary. A component is one multiple-choice question. One possible way to model this component is to generate random digits 0-9. Let digits 0-7 represent a correct answer, and let digits 8 and 9 represent an incorrect answer. Each run will consist of 6 random digits. The response variable is whether or not all 6 simulated questions are answered correctly (all 6 digits are 0-7). The total number of successes divided by the total number of runs will be the simulated probability of getting all 6 questions right. According to the simulation, the probability of getting all 6 multiple-choice questions correct is expected to be about 26%.

13. Driving test.

Answers will vary. A component is one drivers test, but this component will be modeled differently, depending on whether or not it is the first test taken. One possible way to model this component is to generate pairs of random digits 00-99. Let 01-34 represent passing the first test and let 35-99 and 00 represent failing the first test. Let 01-72 represent passing a retest, and let 73-99 and 00 represent failing a retest. To simulate one run, generate pairs of random numbers until a pair is generated that represents passing a test. Begin each run using the "first test" representation, and switch to the "retest" representation if failure is indicated on the first simulated test. The response variable is the number of simulated tests required to achieve the first passing test. The total number of simulated tests taken divided by the total number of runs is the simulated average number of tests required to pass. According to the simulation, the number of driving tests required to pass is expected to be about 1.9.

15. Basketball strategy.

Answers will vary. A component is one foul shot. One possible way to model this component would be to generate pairs of random digits 00-99. Let 01-72 represent a made shot, and let 73-99 and 00 represent a missed shot. The response variable is the number of shots made in a "one and one" situation. If the first shot simulated represents a made shot, simulate a second shot. If the first shot simulated represents a miss, the run is over. The simulated average number of points is the total number of simulated points divided by the total number of runs. According to the simulation, the player is expected to score about 1.24 points.

17. Free groceries.

Answers will vary. A component is the selection of one card with the prize indicated. One possible way to model the prize is to generate pairs of random digits 00-99. Let 01-10 represent $200, let 11-20 represent $100, let 21-40 represent $50, and let 41-99 and 00 represent $20. Repeated pairs of digits must be ignored. (For this reason, a simulation in which random digits 0-9 are generated with 0 representing $200, 1 representing $100, etc., is NOT acceptable. Each card must be individually represented.) A run continues until the total simulated prize is greater than $500. The response variable is the number of simulated customers until the payoff is greater than $500. The simulated average number of customers is the total number of simulated customers divided by the number of runs. According to the simulation, about 10.2 winners are expected each week.

19. The family.

Answers will vary. Each child is a component. One way to model the component is to generate random digits 0-9. Let 0-4 represent a boy and let 5-9 represent a girl. A run consists of generating random digits until a child of each gender is simulated. The response variable is the number of children simulated until this happens. The simulated average family size is the number of digits generated in each run divided by the total number of runs. According to the simulation, the expected number of children in the family is about 3.

21. Dice game.

Answers will very. Each roll of the die is a component. One way of modeling this component is to generate random digits 0-9. The digits 1-6 correspond to the numbers on the faces of the die, and digits 7-9 and 0 are ignored. A run consists of generating random numbers until the sum of the numbers is exactly 10. If the sum exceeds 10, the last roll must be ignored and simulated again, but still counted as a roll. The response variable is the number of rolls until the sum is exactly 10. The simulated average number of rolls until this happens is the total number of rolls simulated divided by the number of runs. According to the simulation, expect to roll the die about 7.5 times.

23. The hot hand.

Answers may vary. Each shot is a component. One way of modeling this component is to generate pairs of random digits 00-99. Let 01-65 represent a made shot, and let 66-99 and 00 represent a missed shot. A run consists of 20 simulated shots. The response variable is whether or not the 20 simulated shots contained a run of 6 or more made shots. To find the simulated percentage of games in which the player is expected to have a run of 6 or more made shots, divide the total number of successes by the total number of runs. According to the simulation, the player is expected to make 6 or more shots in a row in about 40% of games. This isn't unusual. The announcer was wrong to characterize her performance as extraordinary.

25. Job Discrimination

Answers may vary. Each person hired is a component. One way of modeling this component is to generate pairs of random digits 00-99. Let 01-10 represent each of the 10 women, and let 11-22 represent each of the 12 men. Ignore 23-99 and 00. A run consists of 3 usable pairs of numbers. Ignore repeated pairs of digits, since the same man or woman cannot be hired more than once. The response variable is whether or not all 3 simulated hires are women. The simulated percentage of the time that 3 women are expected to be hired is the number of successes divided by the number of runs. According to the simulation, the 3 people hired will all be women about 7.8% of time. This seems a bit strange, but not quite strange enough to be evidence of job discrimination.

27. Second team.

Answers will vary. Each player chosen is a component. One way to model this component is to generate random numbers 0-9. Let digits 1-4 represent the four players who are to be chosen. Ignore digits 5-9 and 0. A run consists of generating a sequence of random numbers that represents the order in which the cards where chosen. Since each number represents a person, and people cannot be chosen more than once, ignore repeated numbers. The response variable is whether or not any digit in the generated sequence matches the corresponding digit in the sequence 1234 (or any other sequence of the four numbers, as long as it is determined ahead of time). The simulated percentage of the time this is expected to happen is the total number of successes (times that the sequences have no matching corresponding digits) divided by the number of runs. According to the simulation, all players are expected to be paired with someone other than the person with whom he or she came to the party about 37.5% of the time.

29. Freshmen.

Answers may vary. Each envelope is a component. According to the Normal model, the probability that one student will score over 1200 on the SAT is approximately 10.6%. One way to model this component is to generate triples of random digits 000-999. Let 001-106 represent a student with score over 1200, and let 107-999 and 000 represent a student with score below 1200. A run consists of simulating envelopes until three are found which represent scores above 1200. The response variable is the number of envelopes simulated until three are found that represent scores above 1200. The simulated average number of envelopes required is the total number of envelopes simulated divided by the number of runs. According to the simulation, the number of envelopes required would be about 25.

Chapter 12 – Sample Surveys

1. **Population** – human resources directors of Fortune 500 companies
   Parameter – proportion of directors who don't feel that surveys intruded on their workday
   Sampling Frame – list of HR directors at Fortune 500 companies
   Sample – the 23% of the HR directors who responded
   Method – questionnaire mailed to all (non-random)
   Bias – nonresponse. It is impossible to generalize about the opinions all HR directors because the nonrespondents would likely have a different opinion about the question. HR directors who feel that surveys intrude on their workday would be less likely to respond.

3. **Population** – all U.S. adults
   Parameter – proportion who have used and benefited from alternative medical treatments.
   Sampling Frame – all Consumers Union subscribers
   Sample – those subscribers who responded
   Method – not specified, but probably a questionnaire mailed to all subscribers
   Bias – nonresponse bias, specifically voluntary response bias. Those who respond may have strong feelings one way or another.

5. **Population** – adults
   Parameter – proportion who think drinking and driving is a serious problem
   Sampling Frame – bar patrons
   Sample – every 10th person leaving the bar
   Method – systematic sampling
   Bias – undercoverage. Those interviewed had just left a bar, and may have opinions about drinking and driving that differ from the opinions of the population in general.

7. **Population** – soil around a former waste dump
   Parameter – concentrations of toxic chemicals in the soil
   Sampling Frame – accessible soil around the dump
   Sample – 16 soil samples
   Method – not clear. There is no indication that the samples were selected randomly.
   Bias – possible convenience sample. Since there is no indication of randomization, the samples may have been taken from easily accessible areas. Soil in these areas may be more or less polluted than the soil in general.

9. **Population** – snack food bags
   Parameter – weight of bags, proportion passing inspection
   Sampling Frame – all bags produced each day
   Sample – 10 bags, one from each of 10 randomly selected cases
   Method – multistage sampling. Presumably, they take a simple random sample of 10 cases, followed by a simple random sample of one bag from each case.
   Bias – no indication of bias.

11. Parent opinion, part 1.

a) This is a voluntary response sample. Only those who see the ad, feel strongly about the issue, and have web access will respond.

b) This is cluster sampling, but probably not a good idea. The opinions of parents in one school may not be typical of the opinions of all parents.

c) This is an attempt at a census, and will probably suffer from nonresponse bias.

d) This is stratified sampling. If the follow-up is carried out carefully, the sample should be unbiased.

13. Wording the survey.

a) Responses to these questions will differ. Question 1 will probably get "no" answers, and Question 2 will probably get "yes" answers. This is response bias, based on the wording of the questions.

b) A question with neutral wording might be: "Do you think standardized tests are

15. Phone surveys.

a) A simple random sample is difficult in this case because there is a problem with undercoverage. People with unlisted phone numbers and those without phones are not in the sampling frame. People who are at work, or otherwise away from home, are included in the sampling frame. These people could never be in the sample itself.

b) One possibility is to generate random phone numbers and call at random times, although obviously not in the middle of the night! This would take care of the undercoverage of people at work during the day, as well as people with unlisted phone numbers, although there is still a problem avoiding undercoverage of people without phones.

c) Under the original plan, those families in which one person stays home are more likely to be included. Under the second plan, many more are included. People without phones are still excluded.

d) Follow-up of this type greatly improves the chance that a selected household is included, increasing the reliability of the survey.

e) Random dialers allow people with unlisted phone numbers to be selected, although they may not be the most willing participants. There is a reason that the phone number is unlisted. Time of day will still be an issue, as will people without phones.

17. Arm length.

a) Answers will vary. My arm length is 3 hand widths and 2 finger widths.

b) The parameter estimated by 10 measurements is the true length of your arm. The population is all possible measurements of your arm length.

c) The population is now the arm lengths of your friends. The average now estimates the mean of the arm lengths of your friends.

d) These 10 arm lengths are unlikely to be representative of the community, or the country. Your friends are likely to be of the same age, and not very diverse.

19. Accounting.

a) Assign numbers 001-120 to each order. Generate 10 random numbers 001-120, and select those orders to recheck.

b) The supervisor should perform a stratified sample, randomly checking a certain percentage of each type of sales, retail and wholesale.

21. Quality control.

a) Select three cases at random, then select one jar randomly from each case.

b) Generate three random numbers between 61-80, with no repeats, to select three cases. Then assign each of the jars in the case a number 01-12, and generate one random number for each case to select the three jars, one from each case.

c) This is not a simple random sample, since there are groups of three jars that cannot be the sample. For example, it is impossible for three jars in the same case to be the sample. This would be possible if the sample were a simple random sample.

23. Sampling methods.

a) This method would probably result in undercoverage of those doctors that are not listed in the Yellow Pages. Using the "line listings" seems fair, as long as all doctors are listed, but using the advertisements would not be a typical list of doctors.

b) This method is not appropriate. This cluster sample will probably contain listings for only one or two types of businesses, not a representative cross-section of businesses.

Chapter 13 – Experiments

1. **a)** This is an experiment, since treatments were imposed.
   b) The subjects studied were 30 patients with bipolar disorder.
   c) The experiment has 1 factor (omega-3 fats from fish oil), at 2 levels (high dose of omega-3 fats from fish oil and no omega-3 fats from fish oil).
   d) 1 factor, at 2 levels gives a total of 2 treatments.
   e) The response variable is "improvement", but there is no indication of how the response variable was measured.
   f) There is no information about the design of the experiment.
   g) The experiment is blinded, since the use of a placebo keeps the patients from knowing whether or not they received the omega-3 fats from fish oils. It is not stated whether or not the evaluators of the "improvement" were blind to the treatment, which would make the experiment double-blind.
   h) Although it needs to be replicated, the experiment can determine whether or not omega-3 fats from fish oils cause improvements in patients with bipolar disorder, at least over the short term. The experiment design would be stronger is it were double-blind.

3. **a)** This is an observational study. The researchers are simply studying traits that already exist in the subjects, not imposing new treatments.
   b) This is a prospective study. The subjects were identified first, then traits were observed.
   c) The subjects are roughly 200 men and women with moderately high blood pressure and normal blood pressure. There is no information about the selection method.
   d) The parameters of interest are memory and reaction time.
   e) An observational study has no random assignment, so there is no way to know that high blood pressure caused subjects to do worse on memory and reaction time tests. A lurking variable, such as age or overall health, might have been the cause. The most we can say is that there was an association between blood pressure and scores on memory and reaction time tests in this group, and recommend a controlled experiment to attempt to determine whether or not there is a cause-and-effect relationship.

5. **a)** This is an observational study. The researchers are simply studying traits that already exist in the subjects, not imposing new treatments.
   b) This is a retrospective study. Researchers studied medical records that already existed.
   c) The subjects were 360,000 Swedish men. The selection process is not stated.
   d) The parameter of interest is the percentage of men with kidney cancer among different groups.
   e) There is no random assignment, so a cause-and-effect relationship between weight or high blood pressure and kidney cancer cannot be established.

7. **a)** This is an experiment, since treatments were imposed on randomly assigned groups.
   b) 24 post-menopausal women were the subjects in this experiment.

c) There is 1 factor (type of drink), at 2 levels (alcoholic and non-alcoholic). (Supplemental estrogen is not a factor in the experiment, but rather a blocking variable. The subjects were not given estrogen supplements as part of the experiment.)

d) 1 factor, with 2 levels, is 2 treatments.

e) The response variable is an increase in estrogen level.

f) This experiment utilizes a blocked design. The subjects were blocked by whether or not they used supplemental estrogen. This design reduces variability in the response variable of estrogen level that may be associated with the use of supplemental estrogen.

g) This experiment does not use blinding.

h) This experiment indicates that drinking alcohol leads to increased estrogen level among those taking estrogen supplements.

9. **a)** This is an experiment, since treatments were imposed on randomly assigned groups.

  b) The experimental units were 10 randomly selected garden locations.

  c) There is 1 factor (type of trap), at 2 levels (the different types of trap).

  d) 1 factor, at 2 levels, results in 2 treatments.

  e) The response variable is the number of bugs in each trap.

  f) This experiment incorporates a blocked design. Placing one of each type of trap at each location reduces variability due to location in the garden.

  g) There is no mention of blinding.

  h) This experiment can determine which trap is more effective at catching bugs.

11. **a)** This is an observational study.

  b) The study is retrospective. Results were obtained from pre-existing church records.

  c) The subjects of the study are women in Finland. The data were collected from church records dating 1640 to 1870, but the selection process is unknown.

  d) The parameter of interest is lifespan.

  e) For this group, having sons was associated with a decrease in lifespan of an average of 34 weeks per son, while having daughters was associated with an unspecified increase in lifespan. As there is no random assignment, there is no way to know that having sons caused a decrease in lifespan.

13. **a)** This is an observational study. (Although some might say that the sad movie was "imposed" on the subjects, this was merely a stimulus used to trigger a reaction, not a treatment designed to attempt to influence some response variable. Researchers merely wanted to observe the behavior of two different groups when each was presented with the stimulus.)

  b) The study is prospective. Researchers identified subjects, and then observed them after the sad movie.

  c) The subjects in this study were people with and without depression. The selection process is not stated.

  d) The parameter of interest is crying response to sad situations.

  e) There is no apparent difference in crying response to sad movies for the depressed and nondepressed groups.

15. a) This is an experiment.
   b) The subjects were rats.
   c) There is 1 factor (sleep deprivation), at 4 levels (no deprivation, 6, 12, and 24 hours).
   d) 1 factor, at 4 levels, results in 4 treatments.
   e) The response variable is glycogen level in the brain.
   f) There is no mention of randomness in the design. Hopefully, the rats were randomly assigned to treatment groups.
   g) There is no mention of blinding.
   h) As long as proper randomization was employed, the conclusion could be that rats deprived of sleep have significantly lower glycogen levels and may need sleep to restore that brain energy fuel. Extrapolating to humans would be speculative.

17. a) This is an experiment. Subjects were randomly assigned to treatments.
   b) The subjects were people experiencing migraines.
   c) There are 2 factors (pain reliever and water temperature). The pain reliever factor has 2 levels (pain reliever or placebo), and the water temperature factor has 2 levels (ice water and regular water).
   d) 2 factors, at 2 levels each, results in 4 treatments.
   e) The response variable is the level of pain relief.
   f) The experiment is completely randomized.
   g) The subjects are blinded to the pain reliever factor through the use of a placebo. The subjects are not blinded to the water factor. They will know whether they are drinking ice water or regular water.
   h) The experiment may indicate whether pain reliever alone or in combination with ice water give pain relief, but patients are not blinded to ice water, so the placebo effect may also be the cause of any relief seen due to ice water.

19. Tomatoes.

Answers may vary. Number the tomatoes plants 1 to 24. Use a random number generator to randomly select 24 numbers from 1 to 24 without replication. Assign the tomato plants matching the first 8 numbers to the first group, the second 8 numbers to the second group, and the third group of 8 numbers to the third group.

21. Mozart.

a) The differences in spatial reasoning scores between the students listening to Mozart and the students sitting quietly were more than would have been expected from ordinary sampling variation.

b)

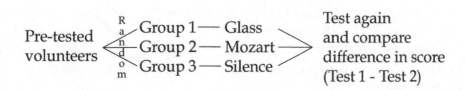

c) The Mozart group seems to have the smallest median difference in spatial reasoning test score and thus the *least* improvement, but there does not appear to be a significant difference.

d) No, the results do not prove that listening to Mozart is beneficial. If anything, there was generally less improvement. The difference does not seem significant compared with the usual variation one would expect between the three groups. Even if type of music has no effect on test score, we would expect some variation between the groups.

23. Frumpies.

a) They should perform a survey. Randomly select a group of children, ages 10 to 13, and have them taste the cereal. Ask the children a question with as little bias as possible, such as, "Do you like the cereal?"

b) Answers may vary. Get volunteers age 10 to 13. Each volunteer will taste both cereals, randomizing the order in which they taste them. Compare the percentage of favorable ratings for each cereal.

Volunteers (10 to 13 years old) — Random — Group 1 — New flavor / Group 2 — Old flavor — Compare percentage of favorable ratings

c) Answers may vary. From the volunteers, identify the children who watch Frump, and identify the children who do not watch Frump. Use a blocked design to reduce variation in cereal preference that may be associated with watching the Frump cartoon.

25. Wine.

a) This is a prospective observational study. The researchers followed a group of children born at a Copenhagen hospital between 1959 and 1961.

b) The results of the Danish study report a link between high socioeconomic status, education, and wine drinking. Since people with high levels of education and higher socioeconomic status are also more likely to be healthy, the relation between health and wine consumption might be explained by the confounding variables of socioeconomic status and education.

c) These studies prove none of these. While the variables have a relation, there is no indication of a cause-and-effect relationship. The only way to determine causation is through a controlled, randomized, and replicated experiment.

27. Dowsing.

a) Arrange the 20 containers in 20 separate locations. Number the containers 01 – 20, and use a random number generator to identify the 10 containers that should be filled with water.

b) We would expect the dowser to be correct about 50% of the time, just by guessing. A record of 60% (12 out of 20) does not appear to be significantly different than the 10 out of 20 expected.

c) Answers may vary. A high level of success would need to be observed. 90% to 100% success (18 to 20 correct identifications) would be convincing.

29. Reading.

Answers may vary. This experiment has 1 factor (reading program), at 2 levels (phonics and whole language), resulting in 2 treatments. The response variable is reading score on an appropriate reading test after a year in the program. After randomly assigning students to teachers, randomly assign half the reading teachers in the district to use each method. There may be variation in reading score based on school within the district, as well as by grade. Blocking by both school and grade will reduce this variation.

31. Weekend deaths.

a) The difference between death rate on the weekend and death rate during the week is greater than would be expected due to natural sampling variation.

b) This was a prospective observational study. The researchers identified hospitals in Ontario, Canada, and tracked admissions to the emergency rooms. This certainly cannot be an experiment. People can't be assigned to become injured on a specific day of the week!

c) Waiting until Monday, if you were ill on Saturday, would be foolish. There are likely to be confounding variables that account for the higher death rate on the weekends. For example, people might be more likely to engage in risky behavior on the weekend.

d) Alcohol use might have something to do with the higher death rate on the weekends. Perhaps more people drink alcohol on weekends, which may lead to more traffic accidents, and higher rates of violence during these days of the week.

33. Beetles.

Answers may vary. This experiment has 1 factor (pesticide), at 3 levels (pesticide A, pesticide B, no pesticide), resulting in 3 treatments. The response variable is the number of beetle larvae found on each plant. Randomly select a third of the plots to be sprayed with pesticide A, a third with pesticide B, and a third to be sprayed with no pesticide (since the researcher also wants to know whether the pesticides even work at all). To control the experiment, the plots of land should be as similar as possible, with regard to amount of sunlight, water, proximity to other plants, etc. If not, plots with similar characteristics should be blocked together. If possible, use some inert substance as a placebo pesticide on the control group, and do not tell the counters of the beetle larvae which plants have been treated with pesticides. After a given period of time, count the number of beetle larvae on each plant and compare the results.

35. Safety switch.

Answers may vary. This experiment has 1 factor (hand), at 2 levels (right, left), resulting in 2 treatments. The response variable is the difference in deactivation time between left and right hand. Find a group of volunteers. Using a matched design, we will require each volunteer to deactivate the machine with his or her left hand, as well as with his or her right hand. Randomly assign the left or right hand to be used first. Hopefully, this will equalize any variability in time that may result from experience gained after deactivating the machine the first time. Complete the first attempt for the whole group. Now repeat the experiment with the alternate hand. Check the differences in time for the left and right hands. Since the response variable is difference in times for each hand, workers should be blocked into groups based on their dominant hand. Another way to account for this difference would be to use the absolute value of the difference as the response variable. We are interested in whether or not the difference is significantly different from the zero difference we would expect if the machine were just as easy to operate with either hand.

Review of Part III – Gathering Data

1. The researchers performed a prospective observational study, since the children were identified at birth and examined at ages 8 and 20. There were indications of behavioral differences between the group of "preemies", and the group of full-term babies. The "preemies" were less likely to engage in risky behaviors, like use of drugs and alcohol, teen pregnancy, and conviction of crimes. This may point to a link between premature birth and behavior, but there may be lurking variables involved. Without a controlled, randomized, and replicated experiment, a cause-and-effect relationship cannot be determined.

3. The researchers at the Purina Pet Institute performed an experiment, matched by gender and weight. The experiment had one factor (diet), at two levels (allowing the dogs to eat as much as they want, or restricted diet), resulting in two treatments. One of each pair of similar puppies was randomly assigned to each treatment. The response variable was length of life. The researchers were able to conclude that, on average, dogs with a lower-calorie diet live longer.

5. It is not apparent whether or not the high folate intake was a randomly imposed treatment. It is likely that the high folate intake was simply an observed trait in the women, so this is a prospective observational study. This study indicates that folate may help in reducing colon cancer for those with family histories of the disease. There may be other variables in the diet and lifestyle of these women that accounted for the decreased risk of colon cancer, so a cause-and-effect relationship cannot be inferred.

7. The fireworks manufacturers are sampling. No information is given about the sampling procedure, so hopefully the tested fireworks are selected randomly. It would probably be a good idea to test a few of each type of firework, so stratification by type seems likely. The population is all fireworks produced each day, and the parameter of interest is the proportion of duds. With a random sample, the manufacturers can make inferences about the proportion of duds in the entire day's production, and use this information to decide whether or not the day's production is suitable for sale.

9. The researchers performed a retrospective observational study. The data were gathered from pre-existing medical records. Living near strong electromagnetic fields may be associated with an increase in leukemia rates. Since this is not a controlled, randomized, and replicated experiment, lurking variables may be involved in this increased risk. For example, the neighborhood around the antennas may consist of people of the same socioeconomic status. Perhaps some variable linked with socioeconomic status may be responsible for the higher leukemia rates.

11. This is an experiment, blocked by sex of the rat. There is one factor (type of hormone), with at least two levels (leptin, insulin). There is a possibility that an additional level of no hormone (or placebo injection) was used as a control, although this is not specifically mentioned. This results in at least two treatments, possibly three, if a control was used. There is no specific mention of random allocation. There are two response variables: amount of food consumed and weight lost. Hormones can help suppress appetite, and

lead to weight loss, in laboratory rats. The male rats responded best to insulin, and the female rats responded best to leptin.

13. The researchers performed an experiment. There is one factor (gene therapy), at two levels (gene therapy and no gene therapy), resulting in two treatments. The experiment is completely randomized. The response variable is heart muscle condition. The researchers can conclude that gene therapy is responsible for stabilizing heart muscle in laboratory rats.

15. The orange juice plant depends on sampling to ensure the oranges are suitable for juice. The population is all of the oranges on the truck, and the parameter of interest is the proportion of unsuitable oranges. The procedure used is a random sample, stratified by location in the truck. Using this well-chosen sample, the workers at the plant can estimate the proportion of unsuitable oranges on the truck, and decide whether or not to accept the load.

17. The researchers performed a prospective observational study, since the subjects were identified ahead of time. Physically fit men may have a lower risk of death from cancer than other men.

19. Point spread.

Answers may vary. Perform a simulation to determine the gambler's expected winnings. A component is one game. To model that component, generate random digits 0 to 9. Since the outcome after the point spread is a tossup, let digits 0-4 represent a loss, and let digits 5-9 represent a win. A run consists of 5 games, so generate 5 random digits at a time. The response variable is the profit the gambler makes, after accounting for the $10 bet. If the outcome of the run is 0, 1, or 2 simulated wins, the profit is −$10. If the outcome is 3, 4, or 5 simulated wins, the profit is $0, $10, or $40, respectively. The total profit divided by the number of runs is the average weekly profit. According to the simulation (80 runs were performed), the gambler is expected to break even. His simulated losses equaled his simulated winnings.

21. Everyday randomness.

Answers will vary. Most of the time, events described as "random" are anything but truly random.

23. Tips.

a) The waiters performed an experiment, since treatments were imposed on randomly assigned groups. This experiment has one factor (candy), at two levels (candy or no candy), resulting in two treatments. The response variable is the percentage of the bill given as a tip.

b) If the decision whether to give candy or not was made before the people were served, the server may have subconsciously introduced bias by treating the customers better. If the decision was made just before the check was delivered, then it is reasonable to conclude that the candy was the cause of the increase in the percentage of the bill given as a tip.

c) "Statistically significant" means that the difference in the percentage of tips between the candy and no candy groups was more than expected due to sampling variability.

25. Cloning.

a) The *USA Weekend* survey suffers from voluntary response bias. Only those who feel strongly will pay for the 900 number phone call.

b) Answers may vary. A strong positive response might be created with the question, "If it would help future generations live longer, healthier lives, would you be in favor of human cloning?"

27. When to stop?

a) Answers may vary. A component in this simulation is rolling 1 die. To simulate this component, generate a random digit 1 to 6. To simulate a run, simulate 4 rolls, stopping if a 6 is rolled. The response variable is the sum of the 4 rolls, or 0 if a 6 is rolled. The average number of points scored is the sum of all rolls divided by the total number of runs. According to the simulation, the average number of points scored will be about 5.8.

b) Answers may vary. A component in this simulation is rolling 1 die. To simulate this component, generate a random digit 1 to 6. To simulate a run, generate random digits until the sum of the digits is at least 12, or until a 6 is rolled. The response variable is the sum of the digits, or 0 if a 6 is rolled. The average number of points scored is the sum of all rolls divided by the total number of runs. According to the simulation, the average number of points scored will be about 5.8, similar to the outcome of the method described in part a).

c) Answers may vary. Be careful when making your decision about the effectiveness of your strategy. If you develop a strategy with a higher simulated average number of points than the other two methods, this is only an indication that you may win in the long run. If the game is played round by round, with the winner of a particular round being declared as the player with the highest roll made during that round, the game is much more variable. For example, if Player B rolls a 12 in a particular game, Player A will always lose that game, provided he or she sticks to the strategy. A better way to get a feel for your chances of winning this type of game might be to simulate several rounds, recording whether each player won or lost the round. Then estimate the percentage of the time that each player is expected to win, according to the simulation.

29. Homecoming.

a) Since telephone numbers were generated randomly, every number that could possibly occur in that community had an equal chance of being selected. This method is "better" than using the phone book, because unlisted numbers are also possible. Those community members who deliberately do not list their phone numbers might not consider this method "better"!

b) Although this method results in a simple random sample of phone numbers, it does not result in a simple random sample of residences. Residences without a phone are excluded, and residences with more than one phone have a greater chance of being included.

c) No, this is not a SRS of local voters. People who respond to the survey may be of the desired age, but not registered to vote. Additionally, some voters who are contacted may choose not to participate.

d) This method does not guarantee an unbiased sample of households. Households in which someone answered the phone may be more likely to have someone at home when the phone call was generated. The attitude about homecoming of these households might not be the same as the attitudes of the community at large.

31. Smoking and Alzheimer's.

a) The studies do not prove that smoking offers any protection from Alzheimer's. The studies merely indicate an association. There may be other variables that can account for this association.

b) Alzheimer's usually shows up late in life. Since smoking is known to be harmful, perhaps smokers have died of other causes before Alzheimer's can be seen.

c) The only way to establish a cause-and-effect relationship between smoking and Alzheimer's is to perform a controlled, randomized, and replicated experiment. This is unlikely to ever happen, since the factor being studied, smoking, has already been proven harmful. It would be unethical to impose this treatment on people for the purposes of this experiment. A prospective observational study could be designed in which groups of smokers and nonsmokers are followed for many years and the incidence of Alzheimer's disease is tracked.

33. Sex and violence.

This experiment has one factor (program content), at three levels (violent, sexual, and neutral), resulting in three treatments. The response variable is the number of brand names recalled after watching the program. Numerous subjects will be randomly assigned to see shows with violent, sexual, or neutral content. They will see the same commercials. After the show, they will be interviewed for their recall of brand names in the commercials.

35. Age and party.

a) The number of respondents is roughly the same for each age category. This may indicate a sample stratified by age category, although it may be a simple random sample.

b) 1404 Democrats were surveyed. $\frac{1404}{4002} \approx 35.1\%$ of the people surveyed were Democrats.

c) If this poll is truly representative, this is probably a good estimate of the percentage of voters who are Democrats. However, telephone surveys are likely to systematically exclude those who do not have phones, a group that is likely to be composed of people who cannot afford phones. If socioeconomic status is linked to political affiliation, this might be a problem.

d) The pollsters were probably attempting to determine whether or not political party is associated with age.

37. Save the grapes.

This experiment has one factor (bird control device), at three levels (scarecrow, netting, and no device), resulting in three treatments. Randomly assign different plots in the vineyard to the different treatments, making sure to ensure adequate separation of plots, so that the possible effect of the scarecrow will not be confounded with the other treatments. The response variable to be measured at the end of the season is the proportion of bird-damaged grapes in each plot.

39. Knees.

a) In an experiment, all subjects must be treated as alike as possible. If there were no "placebo surgery", subjects would know that they had not received the operation, and might react differently. Of course, all volunteers for the experiment must be aware of the

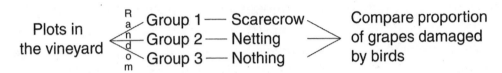

possibility of being randomly assigned to the placebo group.

b) Experiments always use volunteers. This is not a problem, since experiments are testing response to a treatment, not attempting to determine an unknown population parameter. The randomization in an experiment is random assignment to treatment groups, not random selection from a population. Voluntary response is a problem when sampling, but is not an issue in experimentation..

c) There were differences in the amount of pain relief experienced by the two groups, but these differences were small enough that they could be explained by natural sampling variation, even if surgery were not a factor.

41. Security.

a) To ensure that passengers from first-class, as well as coach, get searched, select 2 passengers from first-class and 12 from coach. Using this stratified random sample, 10% of the first-class passengers are searched, as are 10% of the coach passengers.

b) Answers will vary. Number the passengers alphabetically, with 2-digit numbers. Bergman = 01, Bowman = 02, and so on, ending with Testut = 20. Read the random digits in pairs, ignoring pairs 21 to 99 and 00, and ignoring repeated pairs.

```
65|43|67|11|          27|04|
XX XX XX Fontana      XX Castillo
```

The passengers selected for search from first-class are Fontana and Castillo.

c) Number the passengers alphabetically, with 3 digit numbers, 001 to 120. Use the random number table to generate 3-digit numbers, ignoring numbers 121 to 999 and 000, and ignoring repeated numbers. Search the passengers corresponding to the first 12 valid numbers generated.

43. Par 4.

Answers may vary. A component in this simulation is a shot. Use pairs of random digits 00 to 99 to represent a shot. The way in which this component is simulated depends on the type of shot.

For the first shot, let pairs of digits 01 to 70 represent hitting the fairway, and let pairs of digits 71 to 99, and 00, represent not hitting the fairway.

If the first simulated shot hits the fairway, let 01 to 80 represent landing on the green on the second shot, and let 81 to 99, and 00, represent not landing on the green on the second shot. If the first simulated shot does not hit the fairway, let 01 to 40 represent landing on the green on the second shot, and let 41 to 99, and 00, represent not landing on the green on the second shot.

If the second simulated shot does not land on the green, let 01 to 90 represent landing on the green, and 91 to 99, and 00, represent not landing on the green. Keep simulating shots until the shot lands on the green.

Once on the green, let 01 to 20 represent sinking the putt on the first putt, and let 21 to 99, and 00, represent not sinking the putt on the first putt. If second putts are required, continue simulating putts until a putt goes in, with 01 to 90 representing making the putt, and 91 to 99, and 00, representing not making the putt.

A run consists of following the guidelines above until the final putt is made. The response variable is the number of shots required until the final putt is made.

The simulated average score on the hole is the total number of shots required divided by the total number of runs. According to 40 runs of this simulation, a pretty good golfer can be expected to average about 4.2 strokes per hole. Your simulation results may vary.

Chapter 14 – From Randomness to Probability

1. **Roulette.**

   If a roulette wheel is to be considered truly random, then each outcome is equally likely to occur, and knowing one outcome will not affect the probability of the next. Additionally, there is an implication that the outcome is not determined through the use of an electronic random number generator.

3. **Winter.**

   Although acknowledging that there is no law of averages, Knox attempts to use the law of averages to predict the severity of the winter. Some winters are harsh and some are mild over the long run, and knowledge of this can help us to develop a long-term probability of having a harsh winter. However, probability does not compensate for odd occurrences in the short term. Suppose that the probability of having a harsh winter is 30%. Even if there are several mild winters in a row, the probability of having a harsh winter is still 30%.

5. **Cold streak.**

   There is no such thing as being "due for a hit". This statement is based on the so-called law of averages, which is a mistaken belief that probability will compensate in the short term for odd occurrences in the past. The batter's chance for a hit does not change based on recent successes or failures.

7. **Fire insurance.**

   a) It would be foolish to insure your neighbor's house for $300. Although you would probably simply collect $300, there is a chance you could end up paying much more than $300. That risk probably is not worth the $300.

   b) The insurance company insures many people. The overwhelming majority of customers pay the insurance and never have a claim. The few customers who do have a claim are offset by the many who simply send their premiums without a claim. The relative risk to the insurance company is low.

9. **Spinner.**

   a) This is a legitimate probability assignment. Each outcome has probability between 0 and 1, inclusive, and the sum of the probabilities is 1.

   b) This is a legitimate probability assignment. Each outcome has probability between 0 and 1, inclusive, and the sum of the probabilities is 1.

   c) This is not a legitimate probability assignment. Although each outcome has probability between 0 and 1, inclusive, the sum of the probabilities is greater than 1.

   d) This is a legitimate probability assignment. Each outcome has probability between 0 and 1, inclusive, and the sum of the probabilities is 1. However, this game is not very exciting!

   e) This probability assignment is not legitimate. The sum of the probabilities is 0, and there is one probability, –1.5 , that is not between 0 and 1, inclusive.

11. Car repairs.

a) Since all of the events listed are disjoint, the addition rule can be used.

   1. $P(\text{no repairs}) = 1 - P(\text{some repairs}) = 1 - (0.17 + 0.07 + 0.04) = 1 - (0.28) = 0.72$

   2. $P(\text{no more than one repair}) = P(\text{no repairs or one repair}) = 0.72 + 0.17 = 0.89$

   3. $P(\text{some repairs}) = P(\text{one or two or three or more repairs}) = 0.17 + 0.07 + 0.04 = 0.28$

b) Since repairs on the two cars are independent from one another, the multiplication rule can be used. Use the probabilities of events from part a) in the calculations.

   1. $P(\text{neither will need repair}) = (0.72)(0.72) = 0.5184$

   2. $P(\text{both will need repair}) = (0.28)(0.28) = 0.0784$

13. M&M's

a) Since all of the events are disjoint (an M&M can't be two colors at once!), use the addition rule where applicable.

   1. $P(\text{brown}) = 1 - P(\text{not brown}) = 1 - P(\text{yellow or red or orange or blue or green})$
      $= 1 - (0.20 + 0.20 + 0.10 + 0.10 + 0.10) = 0.30$

   2. $P(\text{yellow or orange}) = 0.20 + 0.10 = 0.30$

   3. $P(\text{not green}) = 1 - P(\text{green}) = 1 - 0.10 = 0.90$

   4. $P(\text{striped}) = 0$

b) Since the events are independent (picking out one M&M doesn't affect the outcome of the next pick), the multiplication rule may be used.

   1. $P(\text{all three are brown}) = (0.30)(0.30)(0.30) = 0.027$

   2. $P(\text{the third one is the first one that is red}) = P(\text{not red, not red, red})$
      $= (0.80)(0.80)(0.20) = 0.128$

   3. $P(\text{none are yellow}) = P(\text{not yellow, not yellow, not yellow}) = (0.80)(0.80)(0.80) = 0.512$

   4. $P(\text{at least one is green}) = 1 - P(\text{none are green}) = 1 - (0.90)(0.90)(0.90) = 0.271$

15. Disjoint or independent?

a) For one draw, the events of getting a red M&M and getting an orange M&M are disjoint events. Your single draw cannot be both red and orange.

b) For two draws, the events of getting a red M&M on the first draw and a red M&M on the second draw are independent events. Knowing that the first draw is red does not influence the probability of getting a red M&M on the second draw.

c) Disjoint events can never be independent. Once you know that one of a pair of disjoint events has occurred, the other one cannot occur, so its probability has become zero. For example, consider drawing one M&M. If it is red, it cannot possible be orange. Knowing that the M&M is red influences the probability that the M&M is orange. It's zero. The events are not independent.

17. Dice.

a) $P(6) = \dfrac{1}{6}$, so $P(\text{all 6's}) = \left(\dfrac{1}{6}\right)\left(\dfrac{1}{6}\right)\left(\dfrac{1}{6}\right) \approx 0.005$

b) $P(\text{odd}) = P(1 \text{ or } 3 \text{ or } 5) = \dfrac{3}{6}$, so $P(\text{all odd}) = \left(\dfrac{3}{6}\right)\left(\dfrac{3}{6}\right)\left(\dfrac{3}{6}\right) = 0.125$

c) $P(\text{not divisible by 3}) = P(1 \text{ or } 2 \text{ or } 4 \text{ or } 5) = \dfrac{4}{6}$

   $P(\text{none divisible by 3}) = \left(\dfrac{4}{6}\right)\left(\dfrac{4}{6}\right)\left(\dfrac{4}{6}\right) \approx 0.296$

d) $P(\text{at least one 5}) = 1 - P(\text{no 5's}) = 1 - \left(\dfrac{5}{6}\right)\left(\dfrac{5}{6}\right)\left(\dfrac{5}{6}\right) \approx 0.421$

e) $P(\text{not all 5's}) = 1 - P(\text{all 5's}) = 1 - \left(\dfrac{1}{6}\right)\left(\dfrac{1}{6}\right)\left(\dfrac{1}{6}\right) \approx 0.995$

19. Champion bowler.

Assuming each frame is independent of others, so the multiplication rule may be used.

a) $P(\text{no strikes in 3 frames}) = (0.30)(0.30)(0.30) = 0.027$

b) $P(\text{makes first strike in the third frame}) = (0.30)(0.30)(0.70) = 0.063$

c) $P(\text{at least one strike in the first three frames}) = 1 - P(\text{no strikes}) = 1 - (0.30)^3 = 0.973$

d) $P(\text{perfect game}) = (0.70)^{12} \approx 0.014$

21. Voters.

Since you are calling at random, one person's political affiliation is independent of another's. The multiplication rule may be used.

a) $P(\text{all Republicans}) = (0.29)(0.29)(0.29) \approx 0.024$

b) $P(\text{no Democrats}) = (1 - 0.37)(1 - 0.37)(1 - 0.37) \approx 0.25$

c) $P(\text{at least one Independent}) = 1 - P(\text{no Independents}) = 1 - (0.77)(0.77)(0.77) \approx 0.543$

23. Tires.

Assume that the defective tires are distributed randomly to all tire distributors so that the events can be considered independent. The multiplication rule may be used.

$P(\text{at least one of four tires is defective}) = 1 - P(\text{none are defective})$

$= 1 - (0.98)(0.98)(0.98)(0.98) \approx 0.078$

25. 9/11?

   a) For any date with a valid three-digit date, the chance is 0.001, or 1 in 1000. For many dates in October through December, the probability is 0. For example, there is no way three digits will make 1015, to match October 15.

   b) There are 65 days when the chance to match is 0. (October 10 through October 31, November 10 through November 30, and December 10 through December 31.) That leaves 300 days in a year (that is not a leap year) in which a match might occur.
P(no matches in 300 days) = $(0.999)^{300} \approx 0.741$.

   c) P(at least one match in a year) = $1 - P$(no matches in a year) = $1 - 0.741 \approx 0.259$

   d) P(at least one match on 9/11 in one of the 50 states)
$= 1 - P$(no matches in 50 states) $= 1 - (0.999)^{50} \approx 0.049$

Chapter 15 – Probability Rules!

1. **Sample spaces.**

   a) S = { HH, HT, TH, TT} All of the outcomes are equally likely to occur.

   b) S = { 0, 1, 2, 3} All outcomes are not equally likely. A family of 3 is more likely to have, for example, 2 boys than 3 boys. There are three equally likely outcomes that result in 2 boys (BBG, BGB, and GBB), and only one that results in 3 boys (BBB).

   c) S = { H, TH, TTH, TTT} All outcomes are not equally likely. For example the probability of getting heads on the first try is $\frac{1}{2}$. The probability of getting three tails is $\left(\frac{1}{2}\right)^3 = \frac{1}{8}$.

   d) S = {1, 2, 3, 4, 5, 6} All outcomes are not equally likely. Since you are recording only the larger number of two dice, 6 will be the larger when the other die reads 1, 2, 3, 4, or 5. The outcome 2 will only occur when the other die shows 1 or 2.

3. **Homes.**

   Construct a Venn diagram of the disjoint outcomes.

   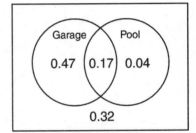

   a) *P*(pool or garage) = *P*(pool) + *P*(garage) – *P*(pool and garage)
   $\quad\quad\quad\quad\quad\quad\quad\quad\;$ = 0.64 + 0.21 – 0.17 = 0.68

   Or, from the Venn: 0.47 + 0.17 + 0.04 = 0.68

   b) *P*(neither)= 1 – *P*(pool or garage) = 1 – 0.68 = 0.32

   Or, from the Venn: 0.32 (the region outside the circles)

   c) *P*(pool and no garage) = *P*(pool) – *P*(pool and garage) = 0.21 – 0.17 = 0.04

   Or, from the Venn: 0.04 (the region inside pool circle, yet outside garage circle)

5. **Amenities.**

   Construct a Venn diagram of the disjoint outcomes.

   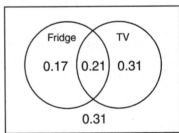

   a) *P*(TV and no refrigerator) = *P*(TV) – *P*(TV and refrigerator)
   $\quad\quad\quad\quad\quad\quad\quad\quad\quad\quad\;$ = 0.52 – 0.21 = 0.31

   Or, from the Venn: 0.31
   (the region inside the TV circle, yet outside the Fridge circle)

   b) *P*(refrigerator or TV, but not both) =
   = [*P*(refrigerator) – *P*(refrigerator and TV)] +[*P*(TV) – *P*(refrigerator and TV)]
   = [0.38 – 0.21] + [0.52 – 0.21] = 0.48

   This problem is much easier to visualize using the Venn diagram. Simply add the probabilities in the two regions for Fridge only and TV only.
   P(refrigerator or TV, but not both) = 0.17 + 0.31 = 0.48

c) $P(\text{neither TV nor refrigerator}) = 1 - P(\text{either TV or refrigerator})$
$$= 1 - [P(TV) + P(\text{refrigerator}) - P(\text{TV and refrigerator})]$$
$$= 1 - [0.52 + 0.38 - 0.21]$$
$$= 0.31$$

Or, from the Venn: 0.31 (the region outside the circles)

7. Cards.

a) $P(\text{heart} \mid \text{red}) = \dfrac{P(\text{heart and red})}{P(\text{red})} = \dfrac{13/52}{26/52} = \dfrac{1}{2}$

A more intuitive approach is to think about only the red cards. Half of them are hearts.

b) $P(\text{red} \mid \text{heart}) = \dfrac{P(\text{red and heart})}{P(\text{heart})} = \dfrac{13/52}{13/52} = 1$

Think about only the hearts. They are all red!

c) $P(\text{ace} \mid \text{red}) = \dfrac{P(\text{ace and red})}{P(\text{red})} = \dfrac{2/52}{26/52} = \dfrac{2}{26} \approx 0.077$

Consider only the red cards. Of those 26 cards, 2 of them are aces.

d) $P(\text{queen} \mid \text{face}) = \dfrac{P(\text{queen and face})}{P(\text{face})} = \dfrac{4/52}{12/52} \approx 0.333$

There are 12 faces cards: 4 jacks, 4 queens, and 4 kings. Four of the 12 face cards are queens.

9. Health.

Construct a two-way table of the conditional probabilities, including the marginal probabilities.

a) $P(\text{both conditions}) = 0.11$

b) $P(\text{high blood pressure}) = 0.11 + 0.16 = 0.27$

		Blood Pressure		
		High	OK	Total
Cholesterol	High	0.11	0.21	0.32
	OK	0.16	0.52	0.68
	Total	0.27	0.73	1.00

c) $P(\text{high chol.} \mid \text{high BP}) = \dfrac{P(\text{high chol. and high BP})}{P(\text{high BP})} = \dfrac{0.11}{0.27} \approx 0.407$

Consider only the High Blood Pressure column. Within this column, the probability of having high cholesterol is 0.11 out of a total of 0.27.

d) $P(\text{high BP} \mid \text{high chol.}) = \dfrac{P(\text{high BP and high chol.})}{P(\text{high chol.})} = \dfrac{0.11}{0.32} \approx 0.344$

This time, consider only the high cholesterol row. Within this row, the probability of having high blood pressure is 0.11, out of a total of 0.32.

11. Sick kids.

Having a fever and having a sore throat are not independent events, so:

P(fever and sore throat) = P(Fever) P(Sore Throat | Fever) = (0.70)(0.30) = 0.21

The probability that a kid with a fever has a sore throat is 0.21.

13. Card.

Since cards are not being replaced in the deck, use conditional probabilities throughout.

a)

P(first heart is drawn on the third card) = P(no heart) P(no heart) P(heart)

$$= \left(\frac{39}{52}\right)\left(\frac{38}{51}\right)\left(\frac{13}{50}\right) \approx 0.145$$

b)

P(all three cards are red) = P(red) P(red) P(red)

$$= \left(\frac{26}{52}\right)\left(\frac{25}{51}\right)\left(\frac{24}{50}\right) \approx 0.118$$

c)

P(none of the cards are spades) = P(no spade) P(no spade) P(no spade)

$$= \left(\frac{39}{52}\right)\left(\frac{38}{51}\right)\left(\frac{37}{50}\right) \approx 0.414$$

d)

P(at least one of the cards in an ace) = $1 - P$(none of the cards are aces)

$$= 1 - \left[P(\text{no ace}) \, P(\text{no ace}) \, P(\text{no ace})\right]$$

$$= 1 - \left(\frac{48}{52}\right)\left(\frac{47}{51}\right)\left(\frac{46}{50}\right) \approx 0.217$$

15. Batteries.

Since batteries are not being replaced, use conditional probabilities throughout.

a)

P(the first two batteries are good) = P(good) P(good)

$$= \left(\frac{7}{12}\right)\left(\frac{6}{11}\right) \approx 0.318$$

b)

P(at least one of the first three batteries works) = $1 - P$(none of the first three batt. work)

$$= 1 - \left[P(\text{no good}) \, P(\text{no good}) \, P(\text{no good})\right]$$

$$= 1 - \left(\frac{5}{12}\right)\left(\frac{4}{11}\right)\left(\frac{3}{10}\right) \approx 0.955$$

c)

$$P(\text{the first four batteries are good}) = P(\text{good})\,P(\text{good})\,P(\text{good})\,P(\text{good})$$

$$= \left(\frac{7}{12}\right)\left(\frac{6}{11}\right)\left(\frac{5}{10}\right)\left(\frac{4}{9}\right) \approx 0.071$$

d)

$$P(\text{pick five to find one good}) = P(\text{no good})\,P(\text{no good})\,P(\text{no good})\,P(\text{no good})\,P(\text{good})$$

$$= \left(\frac{5}{12}\right)\left(\frac{4}{11}\right)\left(\frac{3}{10}\right)\left(\frac{2}{9}\right)\left(\frac{7}{8}\right) \approx 0.009$$

17. Eligibility.

Construct a Venn diagram of the disjoint outcomes.

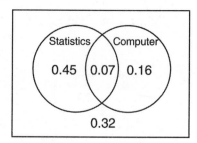

a)

$$P(\text{eligibility}) = P(\text{statistics}) + P(\text{computer science}) - P(\text{both})$$

$$= 0.52 + 0.23 - 0.07$$

$$= 0.68$$

68% of students are eligible for BioResearch, so 100 – 68 = 32% are ineligible.

From the Venn, the region outside the circles represents those students who have taken neither course, and are therefore ineligible for BioResearch.

b)

$$P(\text{computer science} \mid \text{statistics}) = \frac{P(\text{computer science and statistics})}{P(\text{statistics})} = \frac{0.07}{0.52} \approx 0.135$$

From the Venn, consider only the region inside the Statistics circle. The probability of having taken computer science is 0.07 out of a total of 0.52 (the entire Statistics circle).

c) Taking the two courses are not disjoint events, since they have outcomes in common. In fact, 7% of juniors have taken both courses.

d) Taking the two courses are not independent events. The overall probability that a junior has taken a computer science is 0.23. The probability that a junior has taken a computer course given that he or she has taken a statistics course is 0.135. If taking the two courses were independent events, these probabilities would be the same.

19. For sale.

Construct a Venn diagram of the disjoint outcomes.

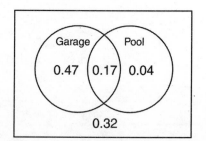

a)

$$P(\text{pool} \mid \text{garage}) = \frac{P(\text{pool and garage})}{P(\text{garage})} = \frac{0.17}{0.64} \approx 0.266$$

From the Venn, consider only the region inside the Garage circle. The probability that the house has a pool is 0.17 out of a total of 0.64 (the entire Garage circle).

b) Having a garage and a pool are not independent events. 26.6% of homes with garages have pools. Overall, 21% of homes have pools. If having a garage and a pool were independent events, these would be the same.

c) No, having a garage and a pool are not disjoint events. 17% of homes have both.

21. Cards.

Yes, getting an ace is independent of the suit when drawing one card from a well shuffled deck. The overall probability of getting an ace is 4/52, or 1/13, since there are 4 aces in the deck. If you consider just one suit, there is only 1 ace out of 13 cards, so the probability of getting an ace given that the card is a diamond, for instance, is 1/13. Since the probabilities are the same, getting an ace is independent of the suit.

23. Men's health, again.

Consider the two-way table from Exercise 9.

High blood pressure and high cholesterol are not independent events. 28.8% of men with OK blood pressure have high cholesterol, while 40.7% of men with high blood pressure have high cholesterol. If having high blood pressure and high cholesterol were independent, these percentages would be the same.

		Blood Pressure		
Cholesterol		High	OK	Total
	High	0.11	0.21	0.32
	OK	0.16	0.52	0.68
	Total	0.27	0.73	1.00

25. Luggage.

Organize the information in a tree diagram.

a) No, the flight leaving on time and the luggage making the connection are not independent events. The probability that the luggage makes the connection is dependent on whether or not the flight is on time. The probability is 0.95 if the flight is on time, and only 0.65 if it is not on time.

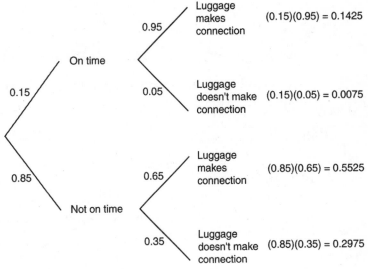

b)
$$P(\text{Luggage}) = P(\text{On time and Luggage}) + P(\text{Not on time and Luggage})$$
$$= (0.15)(0.95) + (0.85)(0.65)$$
$$= 0.695$$

27. Late luggage.

Refer to the tree diagram constructed for Exercise 25.

$$P(\text{Not on time} \mid \text{No Lug.}) = \frac{P(\text{Not on time and No Lug.})}{P(\text{No Lug.})} = \frac{(0.85)(0.35)}{(0.15)(0.05) + (0.85)(0.35)} \approx 0.975$$

If you pick Leah up at the Denver airport and her luggage is not there, the probability that her first flight was delayed is 0.975.

29. Absenteeism.

Organize the information in a tree diagram.

a) No, absenteeism is not independent of shift worked. The rate of absenteeism for the night shift is 2%, while the rate for the day shift is only 1%. If the two were independent, the percentages would be the same.

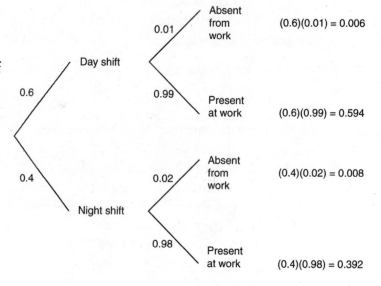

b)

$$P(\text{Absent}) = P(\text{Day and Absent}) + P(\text{Night and Absent}) = (0.6)(0.01) + (0.4)(0.02) = 0.014$$

The overall rate of absenteeism at this company is 1.4%.

31. Absenteeism, part II.

Refer to the tree diagram constructed for Exercise 29.

$$P(\text{Night} \mid \text{Absent}) = \frac{P(\text{Night and Absent})}{P(\text{Absent})} = \frac{(0.4)(0.02)}{(0.6)(0.01) + (0.4)(0.02)} \approx 0.571$$

Approximately 57.1% of the company's absenteeism occurs on the night shift.

33. Drunks.

Organize the information into a tree diagram.

a) $P(\text{Detain} \mid \text{Not Drinking}) = 0.2$

b)

$P(\text{Detain})$

$= P(\text{Drinking and Det.})$

$\quad + P(\text{Not Drinking and Det.})$

$= (0.12)(0.8) + (0.88)(0.2)$

$= 0.272$

c)

$$P(\text{Drunk} \mid \text{Det.}) = \frac{P(\text{Drunk and Det.})}{P(\text{Detain})}$$

$$= \frac{(0.12)(0.8)}{(0.12)(0.8) + (0.88)(0.2)}$$

$$\approx 0.353$$

d)

$$P(\text{Drunk} \mid \text{Release}) = \frac{P(\text{Drunk and Release})}{P(\text{Release})}$$

$$= \frac{(0.12)(0.2)}{(0.12)(0.2) + (0.88)(0.8)}$$

$$\approx 0.033$$

35. Dishwashers.

Organize the information in a tree diagram.

$P(\text{Chuck} \mid \text{Break})$

$$= \frac{P(\text{Chuck and Break})}{P(\text{Break})}$$

$$= \frac{(0.3)(0.03)}{(0.4)(0.01) + (0.3)(0.01) + (0.3)(0.03)}$$

$$\approx 0.563$$

If you hear a dish break, the probability that Chuck is on the job is approximately 0.563.

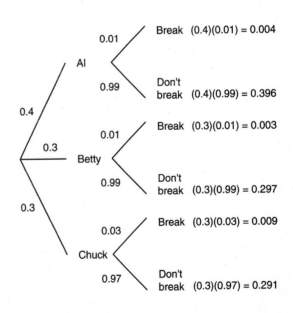

Chapter 16 – Random Variables

1. **Expected value.**

    a) $\mu = E(Y) = 10(0.3) + 20(0.5) + 30(0.2) = 19$

    b) $\mu = E(Y) = 2(0.3) + 4(0.4) + 6(0.2) + 8(0.1) = 4.2$

3. **Pick a card, any card.**

    a)

Win	$0	$5	$10	$30
P(amount won)	$\dfrac{26}{52}$	$\dfrac{13}{52}$	$\dfrac{12}{52}$	$\dfrac{1}{52}$

    b) $\mu = E(\text{amount won}) = \$0\left(\dfrac{26}{52}\right) + \$5\left(\dfrac{13}{52}\right) + \$10\left(\dfrac{12}{52}\right) + \$30\left(\dfrac{1}{52}\right) \approx \$4.13$

    c) Answers may vary. In the long run, the expected payoff of this game is $4.13 per play. Any amount less than $4.13 would be a reasonable amount to pay in order to play. Your decision should depend on how long you intend to play. If you are only going to play a few times, you should risk less.

5. **Kids.**

    a)

Kids	1	2	3
P(Kids)	0.5	0.25	0.25

    b) $\mu = E(\text{Kids}) = 1(0.5) + 2(0.25) + 3(0.25) = 1.75$ kids

    c)

Boys	0	1	2	3
P(boys)	0.5	0.25	0.125	0.125

    $\mu = E(\text{Boys}) = 0(0.5) + 1(0.25) + 2(0.125) + 3(0.125) = 0.875$ boys

7. **Software.**

    Since the contracts are awarded independently, the probability that the company will get both contracts is $(0.3)(0.6) = 0.18$. Organize the disjoint events in a Venn diagram.

Profit	larger only $50,000	smaller only $20,000	both $70,000	neither $0
P(profit)	0.12	0.42	0.18	0.28

    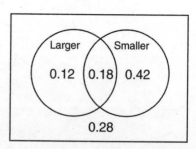

    $\mu = E(\text{profit}) = \$50,000(0.12) + \$20,000(0.42) + \$70,000(0.18)$
    $= \$27,000$

9. **Variation 1.**

   a)
   $$\sigma^2 = Var(Y) = (10-19)^2(0.3) + (20-19)^2(0.5) + (30-19)^2(0.2) = 49$$
   $$\sigma = SD(Y) = \sqrt{Var(Y)} = \sqrt{49} = 7$$

   b)
   $$\sigma^2 = Var(Y) = (2-4.2)^2(0.3) + (4-4.2)^2(0.4) + (6-4.2)^2(0.2) + (8-4.2)^2(0.1) = 3.56$$
   $$\sigma = SD(Y) = \sqrt{Var(Y)} = \sqrt{3.56} \approx 1.89$$

11. **Pick another card.**

   Answers may vary slightly (due to rounding of the mean)
   $$\sigma^2 = Var(\text{Won}) = (0-4.13)^2\left(\frac{26}{52}\right) + (5-4.13)^2\left(\frac{13}{52}\right)$$
   $$+ (10-4.13)^2\left(\frac{12}{52}\right) + (30-4.13)^2\left(\frac{1}{52}\right) \approx 29.5396$$
   $$\sigma = SD(\text{Won}) = \sqrt{Var(\text{Won})} \approx \sqrt{29.5396} \approx \$5.44$$

13. **Kids.**
   $$\sigma^2 = Var(\text{Kids}) = (1-1.75)^2(0.5) + (2-1.75)^2(0.25) + (3-1.75)^2(0.25) = 0.6875$$
   $$\sigma = SD(\text{Kids}) = \sqrt{Var(\text{Kids})} = \sqrt{0.6875} \approx 0.83 \text{ kids}$$

15. **Repairs.**

   a) $\mu = E(\text{Number of Repair Calls}) = 0(0.1) + 1(0.3) + 2(0.4) + 3(0.2) = 1.7$ calls

   b)
   $$\sigma^2 = Var(\text{Calls}) = (0-1.7)^2(0.1) + (1-1.7)^2(0.3) + (2-1.7)^2(0.4) + (3-1.7)^2(0.2) = 0.81$$
   $$\sigma = SD(\text{Calls}) = \sqrt{Var(\text{Calls})} = \sqrt{0.81} = 0.9 \text{ calls}$$

17. **Defects.**

   The percentage of cars with *no* defects is 61%.
   $$\mu = E(\text{Defects}) = 0(0.61) + 1(0.21) + 2(0.11) + 3(0.07) = 0.64 \text{ defects}$$
   $$\sigma^2 = Var(\text{Defects}) = (0-0.64)^2(0.61) + (1-0.64)^2(0.21)$$
   $$+ (2-0.64)^2(0.11) + (3-0.64)^2(0.07) \approx 0.8704$$
   $$\sigma = SD(\text{Defects}) = \sqrt{Var(\text{Defects})} \approx \sqrt{0.8704} \approx 0.93 \text{ defects}$$

19. **Contest.**

   a) The two games are not independent. The probability that you win the second depends on whether or not you win the first.

b)

$$P(\text{losing both games}) = P(\text{losing the first})\, P(\text{losing the second} \mid \text{first was lost})$$
$$= (0.6)(0.7)$$
$$= 0.42$$

c)

$$P(\text{winning both games}) = P(\text{winning the first})\, P(\text{winning the second} \mid \text{first was won})$$
$$= (0.4)(0.2) = 0.08$$

d)

X	0	1	2
$P(X = x)$	0.42	0.50	0.08

e)

$$\mu = E(X) = 0(0.42) + 1(0.50) + 2(0.08) = 0.66 \text{ games}$$
$$\sigma^2 = Var(X) = (0-0.66)^2(0.42) + (1-0.66)^2(0.50) + (2-0.66)^2(0.08) = 0.3844$$
$$\sigma = SD(X) = \sqrt{Var(X)} = \sqrt{0.3844} = 0.62 \text{ games}$$

21. Batteries.

a)

Number good	0	1	2
$P(\text{number good})$	$\left(\dfrac{3}{10}\right)\left(\dfrac{2}{9}\right) = \dfrac{6}{90}$	$\left(\dfrac{3}{10}\right)\left(\dfrac{7}{9}\right) + \left(\dfrac{7}{10}\right)\left(\dfrac{3}{9}\right) = \dfrac{42}{90}$	$\left(\dfrac{7}{10}\right)\left(\dfrac{6}{9}\right) = \dfrac{42}{90}$

b)

$$\mu = E(\text{number good}) = 0\left(\frac{6}{90}\right) + 1\left(\frac{42}{90}\right) + 2\left(\frac{42}{90}\right) = 1.4 \text{ batteries}$$

c)

$$\sigma^2 = Var(\text{number good}) = (0-1.4)^2\left(\frac{6}{90}\right) + (1-1.4)^2\left(\frac{42}{90}\right) + (2-1.4)^2\left(\frac{42}{90}\right) \approx 0.3733$$
$$\sigma = SD(\text{number good}) = \sqrt{Var(\text{number good})} \approx \sqrt{0.3733} \approx 0.61$$

23. Random variables.

a)
$$\mu = E(3X) = 3(E(X)) = 3(10) = 30$$
$$\sigma = SD(3X) = 3(SD(X)) = 3(2) = 6$$

b)
$$\mu = E(Y+6) = E(Y)+6 = 20+6 = 26$$
$$\sigma = SD(Y+6) = SD(Y) = 5$$

c)
$$\mu = E(X+Y) = E(X)+E(Y) = 10+20 = 30$$
$$\sigma = SD(X+Y) = \sqrt{Var(X)+Var(Y)}$$
$$= \sqrt{2^2+5^2} \approx 5.39$$

d)
$$\mu = E(X-Y) = E(X)-E(Y) = 10-20 = -10$$
$$\sigma = SD(X-Y) = \sqrt{Var(X)+Var(Y)}$$
$$= \sqrt{2^2+5^2} \approx 5.39$$

25. Random variables.

a)
$$\mu = E(0.8Y) = 0.8(E(Y)) = 0.8(300) = 240$$
$$\sigma = SD(0.8Y) = 0.8(SD(Y)) = 0.8(16) = 12.8$$

b)
$$\mu = E(2X - 100) = 2(E(X)) - 100 = 140$$
$$\sigma = SD(2X - 100) = 2(SD(X)) = 2(12) = 24$$

c)
$$\mu = E(X + 2Y) = E(X) + 2(E(Y))$$
$$= 120 + 2(300) = 720$$
$$\sigma = SD(X + 2Y) = \sqrt{Var(X) + 2^2 Var(Y)}$$
$$= \sqrt{12^2 + 2^2(16^2)} \approx 34.18$$

d)
$$\mu = E(3X - Y) = 3(E(X)) - E(Y)$$
$$= 3(120) - 300 = 60$$
$$\sigma = SD(3X - Y) = \sqrt{3^2 Var(X) + Var(Y)}$$
$$= \sqrt{3^2(12^2) + 16^2} \approx 39.40$$

27. Eggs.

a) $\mu = E(\text{Broken eggs in 3 dozen}) = 3(E(\text{Broken eggs in 1 dozen})) = 3(0.6) = 1.8$ eggs

b) $\sigma = SD(\text{Broken eggs in 3 dozen}) = \sqrt{0.5^2 + 0.5^2 + 0.5^2} \approx 0.87$ eggs

c) The cartons of eggs must be independent of each other.

29. Repair calls.

$$\mu = E(\text{calls in 8 hours}) = 8(E(\text{calls in 1 hour}) = 8(1.7) = 13.6 \text{ calls}$$
$$\sigma = SD(\text{calls in 8 hours}) = \sqrt{8(Var(\text{calls in 1 hour}))} = \sqrt{8(0.9)^2} \approx 2.55 \text{ hours}$$

This is only valid if the hours are independent of one another.

31. Fire!

a) The standard deviation is large because the profits on insurance are highly variable. Although there will be many small gains, there will occasionally be large losses, when the insurance company has to pay a claim.

b)
$$\mu = E(\text{two policies}) = 2(E(\text{one policy})) = 2(150) = \$300$$
$$\sigma = SD(\text{two policies}) = \sqrt{2(Var(\text{one policy}))} = \sqrt{2(6000^2)} \approx \$8,485.28$$

c)
$$\mu = E(10,000 \text{ policies}) = 10,000(E(\text{one policy})) = 10,000(150) = \$1,500,000$$
$$\sigma = SD(10,000 \text{ policies}) = \sqrt{10,000(Var(\text{one policy}))} = \sqrt{10,000(6000^2)} = \$600,000$$

d) If the company sells 10,000 policies, they are likely to be successful. A profit of \$0, is 2.5 standard deviations below the expected profit. This is unlikely to happen. However, if the company sells fewer policies, then the likelihood of turning a profit decreases. In an extreme case, where only two policies are sold, a profit of \$0 is more likely, being only a small fraction of a standard deviation below the mean.

e) This analysis depends on each of the policies being independent from each other. This assumption of independence may be violated if there are many fire insurance claims as a result of a forest fire, or other natural disaster.

33. Cereal.

a) $E(\text{large bowl} - \text{small bowl}) = E(\text{large bowl}) - E(\text{small bowl}) = 2.5 - 1.5 = 1$ ounce

b) $\sigma = SD(\text{large bowl} - \text{small bowl}) = \sqrt{Var(\text{large}) + Var(\text{small})} = \sqrt{0.4^2 + 0.3^2} = 0.5$ ounces

c)

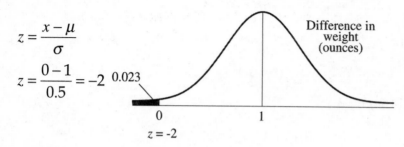

$$z = \frac{x - \mu}{\sigma}$$

$$z = \frac{0 - 1}{0.5} = -2$$

The small bowl will contain more cereal than the large bowl when the difference between the amounts is less than 0. According to the Normal model, the probability of this occurring is approximately 0.023.

d)

$\mu = E(\text{large bowl} + \text{small bowl}) = E(\text{large bowl}) + E(\text{small bowl}) = 2.5 + 1.5 = 4$ ounce

$\sigma = SD(\text{large bowl} + \text{small bowl}) = \sqrt{Var(\text{large}) + Var(\text{small})} = \sqrt{0.4^2 + 0.3^2} = 0.5$ ounces

e)

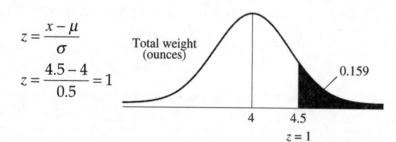

$$z = \frac{x - \mu}{\sigma}$$

$$z = \frac{4.5 - 4}{0.5} = 1$$

According to the Normal model, the probability that the total weight of cereal in the two bowls is more than 4.5 ounces is approximately 0.159.

f)

$\mu = E(\text{box} - \text{large} - \text{small}) = E(\text{box}) - E(\text{large}) - E(\text{small}) = 16.3 - 2.5 - 1.5 = 12.3$ ounces

$\sigma = SD(\text{box} - \text{large} - \text{small}) = \sqrt{Var(\text{box}) + Var(\text{large}) + Var(\text{small})}$

$$= \sqrt{0.2^2 + 0.3^2 + 0.4^2} \approx 0.54 \text{ ounces}$$

35. More cereal.

a) $\mu = E(\text{box} - \text{large} - \text{small}) = E(\text{box}) - E(\text{large}) - E(\text{small}) = 16.2 - 2.5 - 1.5 = 12.2$ ounces

b)

$\sigma = SD(\text{box} - \text{large} - \text{small}) = \sqrt{Var(\text{box}) + Var(\text{large}) + Var(\text{small})}$

$$= \sqrt{0.1^2 + 0.3^2 + 0.4^2} \approx 0.51 \text{ ounces}$$

c)

$$z = \frac{x - \mu}{\sigma}$$

$$z = \frac{13 - 12.2}{0.51}$$

$$z = 1.57$$

Weight left in box (ounces)

0.058

12.2   13

z = 1.57

According to the Normal model, the probability that the box contains more than 13 ounces is about 0.058.

37. Medley.

a)

$$\mu = E(\#1 + \#2 + \#3 + \#4) = E(\#1) + E(\#2) + E(\#3) + E(\#4)$$

$$= 50.72 + 55.51 + 49.43 + 44.91 = 200.57 \text{ seconds}$$

$$\sigma = SD(\#1 + \#2 + \#3 + \#4) = \sqrt{Var(\#1) + Var(\#2) + Var(\#3) + Var(\#4)}$$

$$= \sqrt{0.24^2 + 0.22^2 + 0.25^2 + 0.21^2} \approx 0.46 \text{ seconds}$$

b)

$$z = \frac{x - \mu}{\sigma}$$

$$z = \frac{199.48 - 200.57}{0.461}$$

$$z = -2.36$$

Total time (seconds)

0.009

199.48   200.57

z = -2.36

The team is not likely to swim faster than their best time. According to the Normal model, they are only expected to swim that fast or faster about 0.9% of the time.

39. Farmer's market.

a) Let A = price of a pound of apples, and let P = price of a pound of potatoes.

Profit $= 100A + 50P - 2$

b) $\mu = E(100A + 50P - 2) = 100(E(A)) + 50(E(P)) - 2 = 100(0.5) + 50(0.3) - 2 = \$63$

c) $\sigma = SD(100A + 50P - 2) = \sqrt{100^2(Var(A)) + 50^2(Var(P))} = \sqrt{100^2(0.2^2) + 50^2(0.1^2)} \approx \$20.62$

d) No assumptions are necessary to compute the mean. To compute the standard deviation, independent market prices must be assumed.

Chapter 17 – Probability Models

1. **Bernoulli.**

   a) These are not Bernoulli trials. The possible outcomes are 1, 2, 3, 4, 5, and 6. There are more than two possible outcomes.

   b) These may be considered Bernoulli trials. There are only two possible outcomes, Type A and not Type A. Assuming the 120 donors are representative of the population, the probability of having Type A blood is 43%. The trials are not independent, because the population is finite, but the 120 donors represent less than 10% of the population of all possible donors.

   c) These are not Bernoulli trials. The probability of getting a heart changes as cards are dealt without replacement.

   d) These are not Bernoulli trials. We are sampling without replacement, so the trials are not independent. Samples without replacement may be considered Bernoulli trials if the sample size is less than 10% of the population, but our sample of 500 is more than 10% of 3000.

   e) These may be considered Bernoulli trails. There are only two possible outcomes, sealed properly and not sealed properly. The probability that a package is unsealed is constant, at about 10%, as long as the packages checked are a representative sample of all packages. Finally, the trials are not independent, since the total number of packages is finite, but the 24 packages checked probably represent less than 10% of the packages.

3. **Hoops.**

   The player's shots may be considered Bernoulli trials. There are only two possible outcomes (make or miss), the probability of making a shot is constant (80%), and the shots are independent of one another (making, or missing, a shot does not affect the probability of making the next).

   Let X = the number of shots until the first missed shot.
   Let Y = the number of shots until the first made shot.

   Since these problems deal with shooting until the first miss (or until the first made shot), a geometric model, either $Geom(0.8)$ or $Geom(0.2)$, is appropriate.

   a) Use $Geom(0.2)$. $P(X = 5) = (0.8)^4(0.2) = 0.08192$ (Four shots made, followed by a miss.)

   b) Use $Geom(0.8)$. $P(Y = 4) = (0.2)^3(0.8) = 0.0064$ (Three misses, then a made shot.)

   c) Use $Geom(0.8)$. $P(Y = 1) + P(Y = 2) + P(Y = 3) = (0.8) + (0.2)(0.8) + (0.2)^2(0.8) = 0.992$

5. **More hoops.**

   As determined in Exercise 3, the shots can be considered Bernoulli trials, and since the player is shooting until the first miss, $Geom(0.2)$ is the appropriate model.

   $$E(X) = \frac{1}{p} = \frac{1}{0.2} = 5 \text{ shots}$$ The player is expected to take 5 shots until the first miss.

7. Blood.

These may be considered Bernoulli trials. There are only two possible outcomes, Type AB and not Type AB. Provided that the donors are representative of the population, the probability of having Type AB blood is constant at 4%. The trials are not independent, since the population is finite, but we are selecting fewer than 10% of all potential donors. Since we are selecting people until the first success, the model *Geom*(0.04) may be used.

Let X = the number of donors until the first Type AB donor is found.

a) $E(X) = \dfrac{1}{p} = \dfrac{1}{0.04} = 25$ people   We expect the 25th person to be the first Type AB donor.

b)

P(a Type AB donor among the first 5 people checked)

$= P(X = 1) + P(X = 2) + P(X = 3) + P(X = 4) + P(X = 5)$

$= (0.04) + (0.96)(0.04) + (0.96)^2(0.04) + (0.96)^3(0.04) + (0.96)^4(0.04) \approx 0.185$

c)

P(a Type AB donor among the first 6 people checked)

$= P(X = 1) + P(X = 2) + P(X = 3) + P(X = 4) + P(X = 5) + P(X = 6)$

$= (0.04) + (0.96)(0.04) + (0.96)^2(0.04) + (0.96)^3(0.04) + (0.96)^4(0.04) + (0.96)^5(0.04) \approx 0.217$

d) P(no Type AB donor before the 10th person checked) $= P(X > 9) = (0.96)^9 \approx 0.693$
This one is a bit tricky. There is no implication that we actually find a donor on the 10th trial. We only care that nine trials passed with no Type AB donor.

9. Lefties.

These may be considered Bernoulli trials. There are only two possible outcomes, left-handed and not left-handed. Since people are selected at random, the probability of being left-handed is constant at about 13%. The trails are not independent, since the population is finite, but a sample of 5 people is certainly fewer than 10% of all people.

Let X = the number of people checked until the first lefty is discovered.
Let Y = the number of lefties among $n = 5$.

a) Use *Geom*(0.13).
P(first lefty is the fifth person) $= P(X = 5) = (0.87)^4(0.13) \approx 0.0745$

b) Use *Binom*(5,0.13).
P(some lefties among the 5 people) $= 1 - P$(no lefties among the first 5 people)

$$= 1 - P(Y = 0)$$

$$= 1 - {}_5C_0(0.13)^0(0.87)^5$$

$$\approx 0.502$$

c) Use *Geom*(0.13).
P(first lefty is second or third person) $= P(X = 2) + P(X = 3) = (0.87)(0.13) + (0.87)^2(0.13) \approx 0.211$

d) Use *Binom*(5,0.13).

$P(\text{exactly 3 lefties in the group}) = P(Y = 3) = {}_5C_3(0.13)^3(0.87)^2 \approx 0.0166$

e) Use *Binom*(5,0.13).

$P(\text{at least 3 lefties in the group}) = P(Y = 3) + P(Y = 4) + P(Y = 5)$

$$= {}_5C_3(0.13)^3(0.87)^2 + {}_5C_4(0.13)^4(0.87)^1 + {}_5C_5(0.13)^5(0.87)^0$$
$$\approx 0.0179$$

f) Use *Binom*(5,0.13).

$P(\text{at most 3 lefties in the group}) = P(Y = 0) + P(Y = 1) + P(Y = 2) + P(Y = 3)$

$$= {}_5C_0(0.13)^0(0.87)^5 + {}_5C_1(0.13)^1(0.87)^4$$
$$+ {}_5C_2(0.13)^2(0.87)^3 + {}_5C_3(0.13)^3(0.87)^2$$
$$\approx 0.9987$$

11. Lefties redux.

a) In Exercise 9, we determined that the selection of lefties could be considered Bernoulli trials. Since our group consists of 5 people, use *Binom*(5,0.13).

Let Y = the number of lefties among $n = 5$.

$E(Y) = np = 5(0.13) = 0.65$ lefties

b) $SD(Y) = \sqrt{npq} = \sqrt{5(0.13)(0.87)} \approx 0.75$ lefties

c) Use *Geom*(0.13). Let X = the number of people checked until the first lefty is discovered.

$$E(X) = \frac{1}{p} = \frac{1}{0.13} \approx 7.69 \text{ people}$$

13. Still more lefties.

a) In Exercise 9, we determined that the selection of lefties (and also righties) could be considered Bernoulli trials. Since our group consists of 12 people, and now we are considering the righties, use *Binom*(12,0.87).

Let Y = the number of righties among $n = 12$.

$E(Y) = np = 12(0.87) = 10.44$ righties

$SD(Y) = \sqrt{npq} = \sqrt{12(0.87)(0.13)} \approx 1.16$ righties

b)

$P(\text{not all righties}) = 1 - P(\text{all righties})$

$$= 1 - P(Y = 12)$$
$$= 1 - {}_{12}C_{12}(0.87)^{12}(0.13)^0$$
$$\approx 0.812$$

c)

$P(\text{no more than 10 righties}) = P(Y \le 10)$

$$= P(Y=0) + P(Y=1) + P(Y=2) + \ldots + P(Y=10)$$

$$= {}_{12}C_0(0.87)^0(0.13)^{12} + {}_{12}C_1(0.87)^1(0.13)^{11} + \ldots + {}_{12}C_{10}(0.87)^{10}(0.13)^2$$

$$\approx 0.475$$

d)

$P(\text{exactly six of each}) = P(Y=6)$

$$= {}_{12}C_6(0.87)^6(0.13)^6$$

$$\approx 0.00193$$

e)

$P(\text{majority righties}) = P(Y \ge 7)$

$$= P(Y=7) + P(Y=8) + P(Y=9) + \ldots + P(Y=12)$$

$$= {}_{12}C_7(0.87)^7(0.13)^5 + {}_{12}C_8(0.87)^8(0.13)^4 + \ldots + {}_{12}C_{12}(0.87)^{12}(0.13)^0$$

$$\approx 0.998$$

15. Tennis, anyone?

The first serves can be considered Bernoulli trials. There are only two possible outcomes, successful and unsuccessful. The probability of any first serve being good is given as $p = 0.70$. Finally, we are assuming that each serve is independent of the others. Since she is serving 6 times, use *Binom*(6,0.70).

Let X = the number of successful serves in $n = 6$ first serves.

a)

$P(\text{all six serves in}) = P(X=6)$

$$= {}_6C_6(0.70)^6(0.30)^0$$

$$\approx 0.118$$

b)

$P(\text{exactly four serves in}) = P(X=4)$

$$= {}_6C_4(0.70)^4(0.30)^2$$

$$\approx 0.324$$

c)

$P(\text{at least four serves in}) = P(X=4) + P(X=5) + P(X=6)$

$$= {}_6C_4(0.70)^4(0.30)^2 + {}_6C_5(0.70)^5(0.30)^1 + {}_6C_6(0.70)^6(0.30)^0$$

$$\approx 0.744$$

d)

$P(\text{no more than four serves in}) = P(X=0) + P(X=1) + P(X=2) + P(X=3) + P(X=4)$

$$= {}_6C_0(0.70)^0(0.30)^6 + {}_6C_1(0.70)^1(0.30)^5 + {}_6C_2(0.70)^2(0.30)^4$$

$$+ {}_6C_3(0.70)^3(0.30)^3 + {}_6C_4(0.70)^4(0.30)^2$$

$$\approx 0.580$$

17. And more tennis.

The first serves can be considered Bernoulli trials. There are only two possible outcomes, successful and unsuccessful. The probability of any first serve being good is given as $p = 0.70$. Finally, we are assuming that each serve is independent of the others. Since she is serving 80 times, use *Binom*(80,0.70).

Let X = the number of successful serves in $n = 80$ first serves.

a) $E(X) = np = 80(0.70) = 56$ first serves in.

$SD(X) = \sqrt{npq} = \sqrt{80(0.70)(0.30)} \approx 4.10$ first serves in.

b) Since $np = 56$ and $nq = 24$ are both greater than 10, *Binom*(80,0.70) may be approximated by the Normal model, $N(56, 4.10)$.

c) According to the Normal model, in matches with 80 serves, she is expected to make between 51.9 and 60.1 first serves approximately 68% of the time, between 47.8 and 64.2 first serves approximately 95% of the time, and between 43.7 and 68.3 first serves approximately 99.7% of the time.

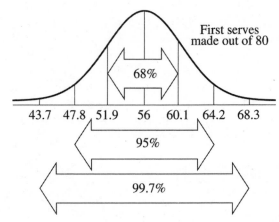

d) **Using *Binom*(80, 0.70):**
$P(\text{at least 65 first serves}) = P(X \geq 65)$

$$= P(X = 65) + P(X = 66) + \ldots + P(X = 80)$$

$$= {}_{80}C_{65}(0.70)^{65}(0.30)^{15} + {}_{80}C_{66}(0.70)^{66}(0.30)^{14} + \ldots + {}_{80}C_{80}(0.70)^{80}(0.30)^{0}$$

$$\approx 0.0161$$

According to the Binomial model, the probability that she makes at least 65 first serves out of 80 is approximately 0.0161.

Using $N(56, 4.10)$:

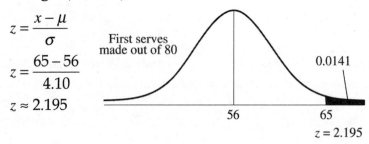

$$z = \frac{x - \mu}{\sigma}$$

$$z = \frac{65 - 56}{4.10}$$

$$z \approx 2.195$$

According to the Normal model, the probability that she makes at least 65 first serves out of 80 is approximately 0.0141.

$P(X \geq 65) \approx P(z > 2.195) \approx 0.0141$

19. Frogs, part II.

The frog examinations can be considered Bernoulli trials. There are only two possible outcomes, having the trait and not having the trait. If the frequency of the trait has not changed, and the biologist collects a representative sample of frogs, then the probability of a frog having the trait is constant, at $p = 0.125$. The trials are not independent since the population of frogs is finite, but 150 frogs is fewer than 10% of all frogs Since the biologist is collecting 150 frogs, use *Binom*(150,0.125).

Let X = the number of frogs with the trait, from $n = 150$ frogs.

a) $E(X) = np = 150(0.125) = 18.75$ frogs.

$SD(X) = \sqrt{npq} = \sqrt{150(0.125)(0.875)} \approx 4.05$ frogs.

b) Since $np = 18.75$ and $nq = 131.25$ are both greater than 10, *Binom*(200,0.125) may be approximated by the Normal model, $N(18.75, 4.05)$.

c) **Using *Binom*(150, 0.125):**
P(at least 22 frogs) $= P(X \geq 22)$

$$= P(X = 22) + \ldots + P(X = 150)$$

$$= {}_{150}C_{22}(0.125)^{22}(0.875)^{128} + \ldots + {}_{150}C_{150}(0.125)^{150}(0.875)^0$$

$$\approx 0.2433$$

According to the Binomial model, the probability that at least 22 frogs out of 150 have the trait is approximately 0.2433.

Using $N(18.75, 4.05)$:

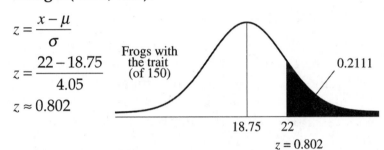

$$z = \frac{x - \mu}{\sigma}$$

$$z = \frac{22 - 18.75}{4.05}$$

$$z \approx 0.802$$

$P(X \geq 22) \approx P(z > 0.802) \approx 0.2111$

According to the Normal model, the probability that at least 22 frogs out of 150 have the trait is approximately 0.2111.

Using either model, the probability that the biologist discovers 22 of 150 frogs with the trait simply as a result of natural variability is quite high. This doesn't prove that the trait has become more common.

21. Lefties again.

Let $X =$ the number of righties among a class of $n = 188$ students.

Using *Binom*(188, 0.87):
These may be considered Bernoulli trials. There are only two possible outcomes, right-handed and not right-handed. The probability of being right-handed is assumed to be constant at about 87%. The trials are not independent, since the population is finite, but a sample of 188 students is certainly fewer than 10% of all people. Therefore, the number of righties in a class of 188 students may be modeled by *Binom*(188, 0.87).

If there are 171 or more righties in the class, some righties have to use a left-handed desk.

P(at least 171 righties) $= P(X \geq 171)$

$$= P(X = 171) + \ldots + P(X = 188)$$

$$= {}_{188}C_{171}(0.87)^{171}(0.13)^{17} + \ldots + {}_{188}C_{188}(0.87)^{188}(0.13)^0$$

$$\approx 0.061$$

According to the binomial model, the probability that a right-handed student has to use a left-handed desk is approximately 0.061.

Using $N(163.56, 4.61)$:

$E(X) = np = 188(0.87) = 163.56$ righties.

$SD(X) = \sqrt{npq} = \sqrt{188(0.87)(0.13)} \approx 4.61$ righties.

Since $np = 163.56$ and $nq = 24.44$ are both greater than 10, $Binom(188, 0.87)$ may be approximated by the Normal model, $N(163.56, 4.61)$.

$z = \dfrac{x - \mu}{\sigma}$

$z = \dfrac{171 - 163.56}{4.61}$

$z \approx 1.614$

Righties in the class of 188

163.56   171

$z = 1.614$

0.0533

$P(X \geq 171) \approx P(z > 1.614) \approx 0.053$

According to the Normal model, the probability that there are at least 171 righties in the class of 188 is approximately 0.0533.

23. Annoying phone calls.

Let X = the number of sales made after making $n = 200$ calls.

Using *Binom*(200, 0.12):

These may be considered Bernoulli trials. There are only two possible outcomes, making a sale and not making a sale. The telemarketer was told that the probability of making a sale is constant at about $p = 0.12$. The trials are not independent, since the population is finite, but 200 calls is fewer than 10% of all calls. Therefore, the number of sales made after making 200 calls may be modeled by $Binom(200, 0.12)$.

$P(\text{at most } 10) = P(X \leq 10)$

$= P(X = 0) + \ldots + P(X = 10)$

$= {}_{200}C_0(0.12)^0(0.88)^{200} + \ldots + {}_{200}C_{10}(0.12)^{10}(0.88)^{190}$

≈ 0.0006

According to the Binomial model, the probability that the telemarketer would make at most 10 sales is approximately 0.0006.

Using *N*(24, 4.60):

$E(X) = np = 200(0.12) = 24$ sales.

$SD(X) = \sqrt{npq} = \sqrt{200(0.12)(0.88)} \approx 4.60$ sales.

Since $np = 24$ and $nq = 176$ are both greater than 10, $Binom(200, 0.12)$ may be approximated by the Normal model, $N(24, 4.60)$.

$$z = \frac{x - \mu}{\sigma}$$

$$z = \frac{10 - 24}{4.60}$$

$$z \approx -3.043$$

0.0012

Sales made in 200 calls

10   24

$z = -3.043$

$$P(X \le 10) \approx P(z < -3.043) \approx 0.0012$$

According to the Normal model, the probability that the telemarketer would make at most 10 sales is approximately 0.0012.

Since the probability that the telemarketer made 10 sales, given that the 12% of calls result in sales is so low, it is likely that he was misled about the true success rate.

25. Seatbelts.

These stops may be considered Bernoulli trials. There are only two possible outcomes, belted or not belted. Police estimate that the probability that a driver is buckled is 80%. (The probability of not being buckled is therefore 20%.) Provided the drivers stopped are representative of all drivers, we can consider the probability constant. The trials are not independent, since the population of drivers is finite, but the police will not stop more than 10% of all drivers.

a) Let X = the number of cars stopped before finding a driver whose seat belt is not buckled. Use *Geom*(0.2) to model the situation.

$$E(X) = \frac{1}{p} = \frac{1}{0.2} = 5 \text{ cars.}$$

b) $P(\text{First unbelted driver is in the sixth car}) = P(X = 6) = (0.8)^5(0.2) \approx 0.066$

c) $P(\text{The first ten drivers are wearing seatbelts}) = (0.8)^{10} \approx .107$

d) Let Y = the number of drivers wearing their seatbelts in 30 cars. Use *Binom*(30, 0.8).

$$E(Y) = np = 30(0.8) = 24 \text{ drivers.}$$

$$SD(Y) = \sqrt{npq} = \sqrt{30(0.8)(0.2)} \approx 2.19 \text{ drivers.}$$

e) Let W = the number of drivers not wearing their seatbelts in 120 cars.

Using *Binom*(120, 0.2):

$P(\text{at least } 20) = P(W \ge 20)$

$\qquad = P(W = 20) + \ldots + P(W = 120)$

$\qquad = {}_{120}C_{20}(0.2)^{20}(0.8)^{100} + \ldots + {}_{120}C_{120}(0.2)^{120}(0.8)^0$

$\qquad \approx 0.848$

According to the Binomial model, the probability that at least 20 out of 120 drivers are not wearing their seatbelts is approximately 0.848.

Using *N*(24, 4.38):

$$E(W) = np = 120(0.2) = 24 \text{ drivers.}$$

$$SD(W) = \sqrt{npq} = \sqrt{120(0.2)(0.8)} \approx 4.38 \text{ drivers.}$$

Since $np = 24$ and $nq = 96$ are both greater than 10, *Binom*(120,0.2) may be approximated by the Normal model, $N(24, 4.38)$.

$$z = \frac{w - \mu}{\sigma}$$

$$z = \frac{20 - 24}{4.38}$$

$$z \approx -0.913$$

Drivers with no seatbelts out of 120 cars

0.8194

20   24

$z = -0.913$

$P(W \geq 120) \approx P(z > -0.913) \approx 0.8194$

According to the Normal model, the probability that at least 20 out of 120 drivers stopped are not wearing their seatbelts is approximately 0.8194.

27. ESP.

Choosing symbols may be considered Bernoulli trials. There are only two possible outcomes, correct or incorrect. Assuming that ESP does not exist, the probability of a correct identification from a randomized deck is constant, at $p = 0.20$. The trials are independent, as long as the deck is shuffled after each attempt. Since 100 trials will be performed, use *Binom*(100, 0.2).

Let X = the number of symbols identified correctly out of 100 cards.

$E(X) = np = 100(0.2) = 20$ correct identifications.

$SD(X) = \sqrt{npq} = \sqrt{100(0.2)(0.8)} = 4$ correct identifications.

Answers may vary. In order be convincing, the "mind reader" would have to identify at least 32 out of 100 cards correctly, since 32 is three standard deviations above the mean. Identifying fewer cards than 32 could happen too often, simply due to chance.

Review of Part IV

1. Quality Control.

Construct a Venn diagram of the disjoint outcomes.

a) $P(\text{defect}) = P(\text{cosm.}) + P(\text{func.}) - P(\text{cosm. and func.})$
$= 0.29 + 0.07 - 0.02 = 0.34$

Or, from the Venn: $0.27 + 0.02 + 0.05 = 0.34$

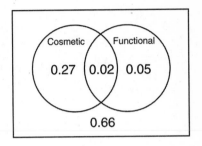

b) $P(\text{cosm. and no func.}) = P(\text{cosm.}) - P(\text{cosm. and func.})$
$= 0.29 - 0.02 = 0.27$

Or, from the Venn: 0.27 (the region inside Cosmetic circle, yet outside Functional circle)

c) $P(\text{func.} \mid \text{cosm.}) = \dfrac{P(\text{func. and cosm.})}{P(\text{cosm.})} = \dfrac{0.02}{0.29} \approx 0.069$

From the Venn, consider only the region inside the Cosmetic circle. The probability that the car has a functional defect is 0.02 out of a total of 0.29 (the entire Cosmetic circle).

d) The two kinds of defects are not disjoint events, since 2% of cars have both kinds.

e) Approximately 6.9% of cars with cosmetic defects also have functional defects. Overall, the probability that a car has a cosmetic defect is 7%. The probabilities are estimates, so these are probably close enough to say that they two types of defects are independent.

3. Airfares.

a) Let C = the price of a ticket to China
Let F = the price of a ticket to France.

Total price of airfare = $3C + 5F$

b) $\mu = E(3C + 5F) = 3E(C) + 5E(F) = 3(1000) + 5(500) = \$5500$

$\sigma = SD(3C + 5F) = \sqrt{3^2(Var(C)) + 5^2(Var(F))} = \sqrt{3^2(150^2) + 5^2(100^2)} \approx \$672.68$

c) $\mu = E(C - F) = E(C) - E(F) = 1000 - 500 = \$500$

$\sigma = SD(C - F) = \sqrt{Var(C) + Var(F)}$
$= \sqrt{150^2 + 100^2} \approx \$180.28$

d) No assumptions are necessary when calculating means. When calculating standard deviations, we must assume that ticket prices are independent of each other for different countries but all tickets to the same country are at the same price.

5. **A game.**

a) Let X = net amount won

X	$0	$2	$-$2
$P(X)$	0.10	0.40	0.50

$\mu = E(X) = 0(0.10) + 2(0.40) - 2(0.50) = -\$0.20$

$\sigma^2 = Var(X) = (0 - (-0.20))^2(0.10) + (2 - (-0.20))^2(0.40) + (-2 - (-0.20))^2(0.50) = 3.56$

$\sigma = SD(X) = \sqrt{Var(X)} = \sqrt{3.56} \approx \$1.89$

b) $X + X$ = the total winnings for two plays.

$\mu = E(X + X) = E(X) + E(X) = (-0.20) + (-0.20) = -\$0.40$

$\sigma = SD(X + X) = \sqrt{Var(X) + Var(X)}$
$= \sqrt{1.89^2 + 1.89^2} \approx \$2.67$

7. **Twins.**

The selection of these women can be considered Bernoulli trials. There are two possible outcomes, twins or no twins. As long as the women selected are representative of the population of all pregnant women, then $p = 1/90$. (If the women selected are representative of the population of women taking Clomid, then $p = 1/10$.) The trials are not independent since the population of all women is finite, but 10 women are fewer than 10% of the population of women.

Let X = the number of twin births from $n = 10$ pregnant women.

Let Y = the number of twin births from $n = 10$ pregnant women taking Clomid.

a) Use $Binom(10, 1/90)$
$P(\text{at least one has twins}) = 1 - P(\text{none have twins})$

$= 1 - P(X = 0)$

$= 1 - {}_{10}C_0 \left(\frac{1}{90}\right)^0 \left(\frac{89}{90}\right)^{10}$

≈ 0.106

b) Use $Binom(10, 1/10)$
$P(\text{at least one has twins}) = 1 - P(\text{none have twins})$

$= 1 - P(Y = 0)$

$= 1 - {}_{10}C_0 \left(\frac{1}{10}\right)^0 \left(\frac{9}{10}\right)^{10}$

≈ 0.651

c) Use *Binom*(5, 1/90) and *Binom*(5, 1/90).

P(at least one has twins) $= 1 - P$(no twins without Clomid)P(no twins with Clomid)

$$= 1 - \left[{}_5C_0\left(\frac{1}{90}\right)^0 \left(\frac{89}{90}\right)^5 \right]\left[{}_5C_0\left(\frac{1}{9}\right)^0 \left(\frac{9}{10}\right)^5 \right]$$

$$\approx 0.442$$

9. More twins.

In Exercise 7, it was determined that these were Bernoulli trials. Use *Binom*(5, 0.10).

Let X = the number of twin births from $n = 5$ pregnant women taking Clomid.

a)

P(none have twins) $= P(X = 0)$

$$= {}_5C_0(0.1)^0(0.9)^5$$

$$\approx 0.590$$

b)

P(exactly one has twins) $= P(X = 1)$

$$= {}_5C_1(0.1)^1(0.9)^4$$

$$\approx 0.328$$

c)

P(at least three will have twins) $= P(X = 3) + P(X = 4) + P(X = 5)$

$$= {}_5C_3(0.1)^3(0.9)^2 + {}_5C_4(0.1)^4(0.9)^1 + {}_5C_5(0.1)^5(0.9)^0$$

$$= 0.00856$$

11. Twins, part III.

In Exercise 7, it was determined that these were Bernoulli trials. Use *Binom*(152, 0.10).

Let X = the number of twin births from $n = 152$ pregnant women taking Clomid.

a) $E(X) = np = 152(0.10) = 15.2$ births.

$SD(X) = \sqrt{npq} = \sqrt{152(0.10)(0.90)} \approx 3.70$ births.

b) Since $np = 15.2$ and $nq = 136.8$ are both greater than 10, the Success/Failure condition is satisfied and *Binom*(152, 0.10) may be approximated by $N(15.2, 3.70)$.

c) Using *Binom*(152, 0.10):

P(no more than 10) $= P(X \leq 10)$

$$= P(X = 0) + \ldots + P(X = 10)$$

$$= {}_{152}C_0(0.10)^0(0.90)^{152} + \ldots + {}_{152}C_{10}(0.10)^{10}(0.90)^{142}$$

$$\approx 0.097$$

According to the Binomial model, the probability that no more than 10 women would have twins is approximately 0.097.

Using N(15.2, 3.70):

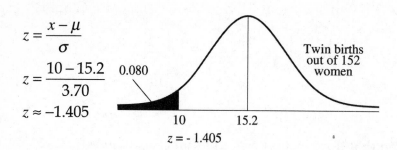

$$z = \frac{x - \mu}{\sigma}$$

$$z = \frac{10 - 15.2}{3.70}$$

$$z \approx -1.405$$

0.080

Twin births out of 152 women

10   15.2

$z = -1.405$

$P(X \le 10) \approx P(z < -1.405) \approx 0.080$

According to the Normal model, the probability that no more than 10 women would have twins is approximately 0.080.

13. Language.

Assuming that the freshman composition class consists of 25 randomly selected people, these may be considered Bernoulli trials. There are only two possible outcomes, having a specified language center or not having the specified language center. The probabilities of the specified language centers are constant at 80%, 10%, or 10%, for right, left, and two-sided language center, respectively. The trials are not independent, since the population of people is finite, but we will select fewer than 10% of all people.

a) Let $L =$ the number of people with left-brain language control from $n = 25$ people.

Use $Binom(25, 0.80)$.

$$P(\text{no more than 15}) = P(L \le 15)$$
$$= P(L = 0) + \ldots + P(L = 15)$$
$$= {}_{25}C_0(0.80)^0(0.20)^{25} + \ldots + {}_{25}C_{15}(0.80)^{15}(0.20)^{10}$$
$$\approx 0.0173$$

According to the Binomial model, the probability that no more than 15 students in a class of 25 will have left-brain language centers is approximately 0.0173.

b) Let $T =$ the number of people with two-sided language control from $n = 5$ people.

Use $Binom(5, 0.10)$.

$$P(\text{none have two - sided language control}) = P(T = 0)$$
$$= {}_5C_0(0.10)^0(0.90)^5$$
$$\approx 0.590$$

c) Use Binomial models:

$E(\text{left}) = np_L = 1200(0.80) = 960$ people

$E(\text{right}) = np_R = 1200(0.10) = 120$ people

$E(\text{two} - \text{sided}) = np_T = 1200(0.10) = 120$ people

d) Let $R =$ the number of people with right-brain language control.

$E(R) = np_R = 1200(0.10) = 120$ people

$SD(R) = \sqrt{np_R q_R} = \sqrt{1200(0.10)(0.90)} \approx 10.39$ people.

e) Since $np_R = 120$ and $nq_R = 1080$ are both greater than 10, the Normal model, $N(120, 10.39)$, may be used to approximate $Binom(1200, 0.10)$. According to the Normal model, about 68% of randomly selected groups of 1200 people could be expected to have between 109.61 and 130.39 people with right-brain language control. About 95% of randomly selected groups of 1200 people could be expected to have between 99.22 and 140.78 people with right-brain language control. About 99.7% of randomly selected groups of 1200 people could be expected to have between 88.83 and 151.17 people with right-brain language control.

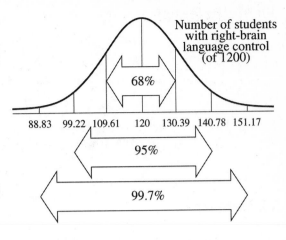

15. Beanstalks.

a) The greater standard deviation for men's heights indicates that men's heights are more variable than women's heights.

b) Admission to a Beanstalk Club is based upon extraordinary height for both men and women, but men are slightly more likely to qualify. The qualifying height for women is about 2.4 standard deviations above the mean height of women, while the qualifying height for men is about 1.75 standard deviations above the mean height for men.

c) Let M = the height of a randomly selected man from $N(69.1, 2.8)$.

Let W = the height of a randomly selected woman from $N(64.0, 2.5)$.

$M - W$ = the difference in height of a randomly selected man and woman.

d) $E(M - W) = E(M) - E(W) = 69.1 - 64.0 = 5.1$ inches

e) $SD(M - W) = \sqrt{Var(M) + Var(W)} = \sqrt{2.8^2 + 2.5^2} \approx 3.75$ inches

f) Since each distribution is described by a Normal model, the distribution of the difference in height between a randomly selected man and woman is $N(5.1, 3.75)$.

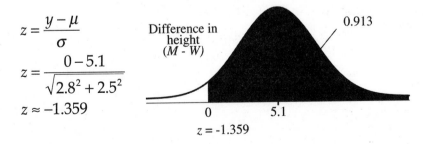

$$z = \frac{y - \mu}{\sigma}$$

$$z = \frac{0 - 5.1}{\sqrt{2.8^2 + 2.5^2}}$$

$$z \approx -1.359$$

According to the Normal model, the probability that a randomly selected man is taller than a randomly selected woman is approximately 0.913.

g) If people chose spouses independent of height, we would expect 91.3% of married couples to consist of a taller husband and shorter wife. The 92% that was seen in the survey is close to 91.3%, and the difference may be due to natural sampling variability. Unless this survey is very large, there is not sufficient evidence of association between height and choice of partner.

17. Multiple choice.

Guessing at questions can be considered Bernoulli trials. There are only two possible outcomes, correct or incorrect. If you are guessing, the probability of success is $p = 0.25$, and the questions are independent. Use $Binom(50, 0.25)$ to model the number of correct guesses on the test.

a) Let $X =$ the number of correct guesses.
$P(\text{at least 30 of 50 correct}) = P(X \geq 30)$
$$= P(X = 30) + \ldots + P(X = 50)$$
$$= {}_{50}C_{30}(0.25)^{30}(0.75)^{20} + \ldots + {}_{50}C_{50}(0.25)^{50}(0.75)^{0}$$
$$\approx 0.00000016$$

You are **very** unlikely to pass by guessing on every question.

b) Use $Binom(50, 0.70)$.
$P(\text{at least 30 of 50 correct}) = P(X \geq 30)$
$$= P(X = 30) + \ldots + P(X = 50)$$
$$= {}_{50}C_{30}(0.70)^{30}(0.30)^{20} + \ldots + {}_{50}C_{50}(0.70)^{50}(0.30)^{0}$$
$$\approx 0.952$$

According to the Binomial model, your chances of passing are about 95.2%.

c) Use $Geom(0.70)$.
$P(\text{first correct on third question}) = (0.30)^{2}(0.70) = 0.063$

19. Insurance.

The company is expected to pay $100,000 only 2.6% of the time, while always gaining $520 from every policy sold. When they pay, they actually only pay $99,480.

$E(\text{profit}) = \$520(0.974) - \$99{,}480(0.026) = -\$2080$.

The expected profit is actually a **loss** of $2080 per policy. The company had better raise its premiums if it hopes to stay in business.

21. Passing stats.

Organize the information in a tree diagram.

a)

$P(\text{Passing Statistics})$
$\quad = P(\text{Scedastic and Pass})$
$\qquad + P(\text{Kurtosis and Pass})$
$\quad \approx 0.4667 + 0.25$
$\quad \approx 0.717$

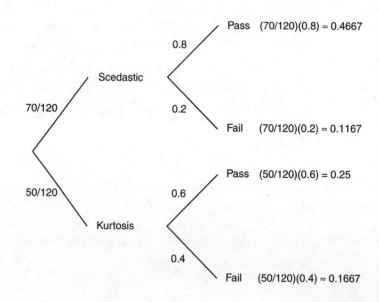

b)

$$P(\text{Kurtosis} \mid \text{Fail})$$

$$= \frac{P(\text{Kurtosis and Fail})}{P(\text{Fail})}$$

$$\approx \frac{0.1667}{0.1167 + 0.1667} \approx 0.588$$

23. Random variables.

a)
$$\mu = E(X+50) = E(X)+50 = 50+50 = 100$$
$$\sigma = SD(X+50) = SD(X) = 8$$

b)
$$\mu = E(10Y) = 10E(Y) = 10(100) = 1000$$
$$\sigma = SD(10Y) = 10SD(Y) = 60$$

c)
$$\mu = E(X+0.5Y) = E(X)+0.5E(Y)$$
$$= 50+0.5(100) = 100$$
$$\sigma = SD(X+0.5Y) = \sqrt{Var(X)+0.5^2 Var(Y)}$$
$$= \sqrt{8^2+0.5^2(6^2)} \approx 8.54$$

d)
$$\mu = E(X-Y) = E(X)-E(Y) = 50-100 = -50$$
$$\sigma = SD(X-Y) = \sqrt{Var(X)+Var(Y)}$$
$$= \sqrt{8^2+6^2} = 10$$

25. Youth survey.

a) Many boys play computer games and use email, so the probabilities can total more than 100%. There is no evidence that there is a mistake in the report.

b) Playing computer games and using email are not disjoint. If they were, the probabilities would total 100% or less.

c) Emailing friends and gender are not independent. 76% of girls emailed friends in the last week, but only 65% of boys emailed. If emailing were independent of gender, the probabilities would be the same.

d) Let X = the number of students chosen until the first student is found who does not use the Internet. Use $Geom(0.07)$. $P(X=5) = (0.93)^4 (0.07) \approx 0.0524$.

27. Travel to Kyrgyzstan.

a) If you spend an average of 4237 soms per day, you can stay about $\frac{90,000}{4237} \approx 21$ days.

b) Assuming that your daily spending is independent, the standard deviation is the square root of the sum of the variances for 21 days.

$$\sigma = \sqrt{21(360)^2} \approx 1649.73 \text{ soms}$$

c) The standard deviation in your total expenditures is about 1650 soms, so if you don't think you will exceed your expectation by more than 2 standard deviations, bring an extra 3300 soms. This gives you a cushion of about 157 soms for each of the 21 days.

29. Home sweet home.

Since the homes are randomly selected, these can be considered Bernoulli trials. There are only two possible outcomes, owning the home or not owning the home. The probability of any randomly selected resident home being owned by the current resident is 0.66. The trials are not independent, since the population is finite, but as long as the city has more than 8200 homes, we are not sampling more than 10% of the population. The Binomial model, *Binom*(820, 0.66), can be used to model the number of homeowners among the 820 homes surveyed. Let H = the number of homeowners found in n = 820 homes.

$E(H) = np = 820(0.66) = 541.2$ homes.

$SD(H) = \sqrt{npq} = \sqrt{820(0.66)(0.34)} \approx 13.56$ homes.

The 523 homeowners found in the candidate's survey represent a number of homeowners that is only about 1.34 standard deviations below the expected number of homeowners. It is not particularly unusual to be 1.34 standard deviations below the mean. There is little support for the candidate's claim of a low level of home ownership.

31. Who's the boss?

a) $P(\text{first three owned by women}) = (0.26)^3 \approx 0.018$

b) $P(\text{none of the first four are owned by women}) = (0.74)^4 \approx 0.300$

c) $P(\text{sixth firm called is owned by women} \mid \text{none of the first five were}) = 0.26$

Since the firms are chosen randomly, the fact that the first five firms were owned by men has no bearing on the ownership of the sixth firm.

33. When to stop?

a) Since there are only two outcomes, 6 or not 6, the probability of getting a 6 is 1/6, and the trials are independent, these are Bernoulli trials. Use *Geom*(1/6).

$$\mu = \frac{1}{p} = \frac{1}{\frac{1}{6}} = 6 \text{ rolls}$$

b) If 6's are not allowed, the mean of each die roll is $\frac{1+2+3+4+5}{5} = 3$. You would expect to get 15 if you rolled 5 times.

c) $P(\text{5 rolls without a 6}) = \left(\frac{5}{6}\right)^5 \approx 0.402$

35. Technology on campus.

Construct a Venn diagram of the disjoint outcomes.

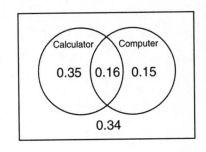

a)

$$P(\text{neither tech.}) = 1 - P(\text{either tech.})$$
$$= 1 - \left[P(\text{calculator}) + P(\text{computer}) - P(\text{both}) \right]$$
$$= 1 - \left[0.51 + 0.31 - 0.16 \right]$$
$$= 0.34$$

Or, from the Venn: 0.34 (the region outside both circles) This is MUCH easier.

34% of students use neither type of technology.

b) $P(\text{calc. and no comp.}) = P(\text{calc.}) - P(\text{calc. and comp.}) = 0.51 - 0.16 = 0.35$

Or, from the Venn: 0.35 (region inside the Calculator circle, outside the Computer circle)

35% of students use calculators, but not computers.

c) $P(\text{computer} \mid \text{calculator}) = \dfrac{P(\text{comp. and calc.})}{P(\text{calc.})} = \dfrac{0.16}{0.51} \approx 0.314$

About 31.4% of calculator users have computer assignments.

d) The percentage of computer users overall is 31%, while 31.4% of calculator users were computer users. These are very close. There is no indication of an association between computer use and calculator use.

37. Socks.

Since we are sampling without replacement, use conditional probabilities throughout.

a) $P(2 \text{ blue}) = \left(\dfrac{4}{12} \right)\left(\dfrac{3}{11} \right) = \dfrac{12}{132} = \dfrac{1}{11}$

b) $P(\text{no grey}) = \left(\dfrac{7}{12} \right)\left(\dfrac{6}{11} \right) = \dfrac{42}{132} = \dfrac{7}{22}$

c) $P(\text{at least one black}) = 1 - P(\text{no black}) = 1 - \left(\dfrac{9}{12} \right)\left(\dfrac{8}{11} \right) = \dfrac{60}{132} = \dfrac{5}{11}$

d) $P(\text{green}) = 0$ (There aren't any green socks in the drawer.)

e) $P(\text{match}) = P(2 \text{ blue}) + P(2 \text{ grey}) + P(2 \text{ black}) = \left(\dfrac{4}{12} \right)\left(\dfrac{3}{11} \right) + \left(\dfrac{5}{12} \right)\left(\dfrac{4}{11} \right) + \left(\dfrac{3}{12} \right)\left(\dfrac{2}{11} \right) = \dfrac{38}{132} = \dfrac{19}{66}$

39. The Drake equation.

a) $N \cdot f_p$ represents the number of stars in the Milky Way Galaxy expected to have planets.

b) $N \cdot f_p \cdot n_e \cdot f_i$ represents the number of planets in the Milky Way Galaxy expected to have intelligent life.

c) $f_l \cdot f_i$ is the probability that a planet has a suitable environment and has intelligent life.

d) $f_l = P(\text{life} \mid \text{suitable environment})$. This is the probability that life develops, if a planet has a suitable environment.

   $f_i = P(\text{intelligence} \mid \text{life})$. This is the probability that the life develops intelligence, if a planet already has life.

   $f_c = P(\text{communication} \mid \text{intelligence})$. This is the probability that radio communication develops, if a planet already has intelligent life.

41. Pregnant?

Organize the information in a tree diagram.

$P(\text{pregnant} \mid \text{positive test})$

$= \dfrac{P(\text{pregnant and positive test})}{P(\text{positive test})}$

$= \dfrac{0.686}{0.686 + 0.006}$

≈ 0.991

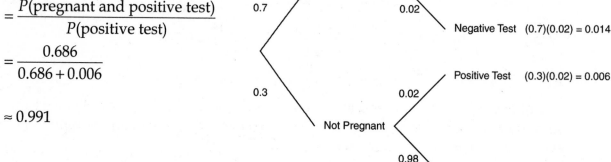

Chapter 18 – Sampling Distribution Models

1. **Coin tosses.**

   a) The histogram of these proportions is expected to be symmetric, but **not** because of the Central Limit Theorem. The sample of 16 coin flips is not large. The distribution of these proportions is expected to be symmetric because the probability that the coin lands heads is the same as the probability that the coin lands tails.

   b) The histogram is expected to have its center at 0.5, the probability that the coin lands heads.

   c) The standard deviation of data displayed in this histogram should be approximately equal to the standard deviation of the sampling distribution model, $\sqrt{\dfrac{pq}{n}} = \sqrt{\dfrac{(0.5)(0.5)}{16}} = 0.125$.

   d) The expected number of heads, $np = 16(0.5) = 8$, which is less than 10. The Success/Failure condition is not met. The Normal model is not appropriate in this case.

3. **More coins.**

   a) $\mu_{\hat{p}} = p = 0.5$ and $\sigma(\hat{p}) = \sqrt{\dfrac{pq}{n}} = \sqrt{\dfrac{(0.5)(0.5)}{25}} = 0.1$

   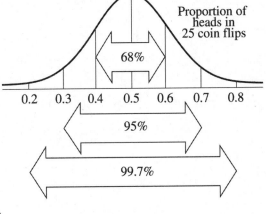

   About 68% of the sample proportions are expected to be between 0.4 and 0.6, about 95% are expected to be between 0.3 and 0.7, and about 99.7% are expected to be between 0.2 and 0.8.

   b) First of all, coin flips are independent of one another. There is no need to check the 10% Condition. Second, $np = nq = 12.5$, so both are greater than 10. The Success/Failure condition is met, so the sampling distribution model is $N(0.5, 0.1)$.

   c) $\mu_{\hat{p}} = p = 0.5$ and $\sigma(\hat{p}) = \sqrt{\dfrac{pq}{n}} = \sqrt{\dfrac{(0.5)(0.5)}{64}} = 0.0625$

   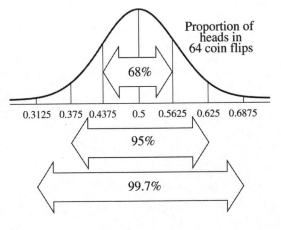

   About 68% of the sample proportions are expected to be between 0.4375 and 0.5625, about 95% are expected to be between 0.375 and 0.625, and about 99.7% are expected to be between 0.3125 and 0.6875.

   Coin flips are independent of one another, and $np = nq = 32$, so both are greater than 10. The Success/Failure condition is met, so the sampling distribution model is $N(0.5, 0.0625)$.

   d) As the number of tosses increases, the sampling distribution model will still be Normal and centered at 0.5, but the standard deviation will decrease. The sampling distribution model will be less spread out.

5. **Just (un)lucky.**

For 200 flips, the sampling distribution model is Normal with $\mu_{\hat{p}} = p = 0.5$ and

$$\sigma(\hat{p}) = \sqrt{\frac{pq}{n}} = \sqrt{\frac{(0.5)(0.5)}{200}} \approx 0.0354.$$ Her sample proportion of $\hat{p} = 0.42$ is about 2.26 standard deviations below the expected proportion, which is unusual, but not extraordinary. According to the Normal model, we expect sample proportions this low or lower about 1.2% of the time.

7. **Speeding.**

a) $\mu_{\hat{p}} = p = 0.70$

$$\sigma(\hat{p}) = \sqrt{\frac{pq}{n}} = \sqrt{\frac{(0.7)(0.3)}{80}} \approx 0.051.$$

About 68% of the sample proportions are expected to be between 0.649 and 0.751, about 95% are expected to be between 0.598 and 0.802, and about 99.7% are expected to be between 0.547 and 0.853.

b) **10% condition:** Although 80 cars represent less than 10% of all cars, the sample may not be representative. If the flow of traffic is very fast, the speed of the other cars around may have some effect on the speed of each driver. Likewise, if traffic is slow, the police may find a smaller proportion of speeders than they expect.

Success/Failure condition: $np = 56$ and $nq = 24$ are both greater than 10.

The Normal model may not be appropriate. Use caution. (And don't speed!)

9. **Loans.**

a) $\mu_{\hat{p}} = p = 7\%$

$$\sigma(\hat{p}) = \sqrt{\frac{pq}{n}} = \sqrt{\frac{(0.07)(0.93)}{200}} \approx 1.8\%$$

b) **10% condition:** Assume that the 200 people are a representative sample of all loan recipients. A sample of this size is less than 10% of all loan recipients.

Success/Failure condition: $np = 14$ and $nq = 186$ are both greater than 10.

Therefore, the sampling distribution model for the proportion of 200 loan recipients who will not make payments on time is $N(0.07, 0.018)$.

c) According to the Normal model, the probability that over 10% of these clients will not make timely payments is approximately 0.048.

$$z = \frac{\hat{p} - \mu_{\hat{p}}}{\sqrt{\frac{pq}{n}}}$$

$$z = \frac{0.10 - 0.07}{\sqrt{\frac{(0.07)(0.93)}{200}}}$$

$$z \approx 1.663$$

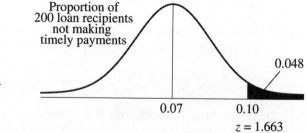

11. Polling.

10% condition: We will assume that the 400 voters polled were chosen randomly, and represent less than 10% of potential voters.

Success/Failure condition: $np = 208$ and $nq = 192$ are both greater than 10.

Therefore, the sampling distribution model for $\hat{p}$ is Normal, with:

$$\mu_{\hat{p}} = p = 0.52$$
$$\sigma(\hat{p}) = \sqrt{\frac{pq}{n}} = \sqrt{\frac{(0.52)(0.48)}{400}} \approx 0.025$$

According to the Normal model, the probability that the newspaper's sample will lead them to predict defeat (that is, predict budget support below 50%) is approximately 0.212.

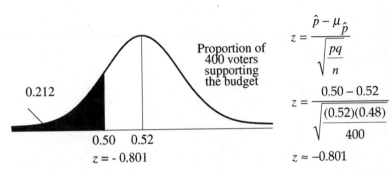

$$z = \frac{\hat{p} - \mu_{\hat{p}}}{\sqrt{\frac{pq}{n}}}$$
$$z = \frac{0.50 - 0.52}{\sqrt{\frac{(0.52)(0.48)}{400}}}$$
$$z \approx -0.801$$

13. Apples.

10% condition: A random sample of 150 apples is taken from each truck, and 150 is less than 10% of all apples.

Success/Failure Condition: $np = 12$ and $nq = 138$ are both greater than 10.

Therefore, the sampling distribution model for $\hat{p}$ is Normal, with:

$$\mu_{\hat{p}} = p = 0.08$$
$$\sigma(\hat{p}) = \sqrt{\frac{pq}{n}} = \sqrt{\frac{(0.08)(0.92)}{150}} \approx 0.0222$$

According to the Normal model, the probability that less than 5% of the apples in the sample are unsatisfactory is approximately 0.088.

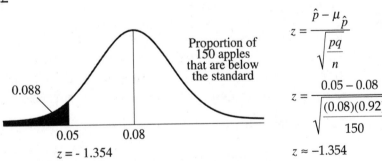

$$z = \frac{\hat{p} - \mu_{\hat{p}}}{\sqrt{\frac{pq}{n}}}$$
$$z = \frac{0.05 - 0.08}{\sqrt{\frac{(0.08)(0.92)}{150}}}$$
$$z \approx -1.354$$

15. Nonsmokers.

10% condition: We will assume that the 120 customers (to fill the restaurant to capacity) are representative of all customers, and represent less than 10% of all potential customers.

Success/Failure condition: $np = 72$ and $nq = 48$ are both greater than 10.

Therefore, the sampling distribution model for $\hat{p}$ is Normal, with:

$$\mu_{\hat{p}} = p = 0.60$$

$$\sigma(\hat{p}) = \sqrt{\frac{pq}{n}} = \sqrt{\frac{(0.60)(0.40)}{120}} \approx 0.0447$$

Answers may vary. We will use 3 standard deviations above the expected proportion of customers who demand nonsmoking seats to be "very sure".

$$\mu_{\hat{p}} + 3\left(\sqrt{\frac{pq}{n}}\right) \approx 0.60 + 3(0.0447) \approx 0.734$$

Since 120(0.734) = 88.08, the restaurant needs at least 89 seats in the nonsmoking section.

17. Sampling.

a) The sampling distribution model for the sample mean is $N\left(\mu, \dfrac{\sigma}{\sqrt{n}}\right)$.

b) If we choose a larger sample, the mean of the sampling distribution model will remain the same, but the standard deviation will be smaller.

19. GPAs.

Random sampling condition: Assume that the students are randomly assigned to seminars.

Independence assumption: It is reasonable to think that GPAs for randomly selected students are mutually independent.

10% condition: The 25 students in the seminar certainly represent less than 10% of the population of students.

The mean GPA for the freshmen was $\mu = 3.4$, with standard deviation $\sigma = 0.35$. Since the distribution of GPAs is roughly unimodal and symmetric, and the sample size is not too small, the Central Limit Theorem tells us that we can model the sampling distribution of the mean GPA with a Normal model, with $\mu_{\bar{y}} = 3.4$ and standard deviation $\sigma(\bar{y}) = \dfrac{0.35}{\sqrt{25}} \approx 0.07$.

The sampling distribution model for the sample mean GPA is approximately $N(3.4, 0.07)$.

21. Pregnancy.

a)

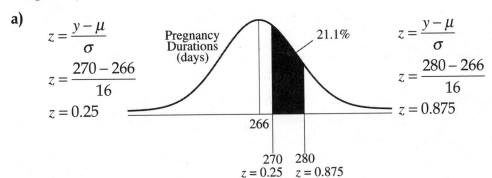

$$z = \frac{y - \mu}{\sigma}$$

$$z = \frac{270 - 266}{16}$$

$$z = 0.25$$

$$z = \frac{y - \mu}{\sigma}$$

$$z = \frac{280 - 266}{16}$$

$$z = 0.875$$

According to the Normal model, approximately 21.1% of all pregnancies are expected to last between 270 and 280 days.

b)

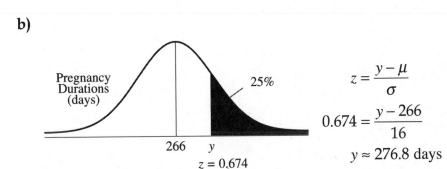

$$z = \frac{y - \mu}{\sigma}$$

$$0.674 = \frac{y - 266}{16}$$

$$y \approx 276.8 \text{ days}$$

According to the Normal model, the longest 25% of pregnancies are expected to last approximately 276.8 days or more.

c) **Random sampling condition:** Assume that the 60 women the doctor is treating can be considered a representative sample of all pregnant women.

Independence assumption: It is reasonable to think that the durations of the patients' pregnancies are mutually independent.

10% condition: The 60 women that the doctor is treating certainly represent less than 10% of the population of all women.

The mean duration of the pregnancies was $\mu = 266$ days, with standard deviation $\sigma = 16$ days. Since the distribution of pregnancy durations is Normal, we can model the sampling distribution of the mean pregnancy duration with a Normal model, with

$\mu_{\bar{y}} = 266$ days and standard deviation $\sigma(\bar{y}) = \dfrac{16}{\sqrt{60}} \approx 2.07$ days.

d) According to the Normal model, the probability that the mean pregnancy duration is less than 260 days is 0.002.

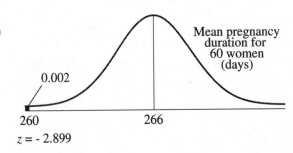

23. Pregnant again.

a) The distribution of pregnancy durations may be skewed to the left since there are more premature births than very long pregnancies. Modern practice of medicine stops pregnancies at about 2 weeks past normal due date by inducing labor or performing a Caesarean section.

b) We can no longer answer the questions posed in parts a and b. The Normal model is not appropriate for skewed distributions. The answer to part c is still valid. The Central Limit Theorem guarantees that the sampling distribution model is Normal when the sample size is large.

25. Dice and dollars.

a) Let X = the number of dollars won in one play.

$$\mu = E(X) = 0\left(\frac{3}{6}\right) + 1\left(\frac{2}{6}\right) + 10\left(\frac{1}{6}\right) = \$2$$

$$\sigma^2 = Var(X) = (0-2)^2\left(\frac{3}{6}\right) + (1-2)^2\left(\frac{2}{6}\right) + (10-2)^2\left(\frac{1}{6}\right) = 13$$

$$\sigma = SD(X) = \sqrt{Var(X)} = \sqrt{13} \approx \$3.61$$

b) $X + X$ = the total winnings for two plays.

$$\mu = E(X + X) = E(X) + E(X) = 2 + 2 = \$4$$

$$\sigma = SD(X + X) = \sqrt{Var(X) + Var(X)}$$
$$= \sqrt{13 + 13} \approx \$5.10$$

c) In order to win at least \$100 in 40 plays, you must average at least $\frac{100}{40} = \$2.50$ per play.

The expected value of the winnings is $\mu = \$2$, with standard deviation $\sigma = \$3.61$. Rolling a die is random and the outcomes are mutually independent, so the Central Limit Theorem guarantees that the sampling distribution model is Normal with with $\mu_{\bar{x}} = \$2$ and standard deviation $\sigma(\bar{x}) = \frac{\$3.61}{\sqrt{40}} \approx \$0.571$.

According to the Normal model, the probability that you win at least \$100 in 40 plays is approximately 0.191.

(This is equivalent to using $N(80, 22.83)$ to model your total winnings.)

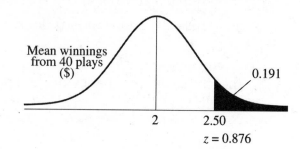

27. Pollution.

a) **Random sampling condition:** Assume that the 80 cars can be considered a representative sample of all cars of this type.

   Independence assumption: It is reasonable to think that the CO emissions for these cars are mutually independent.

   10% condition: The 80 cars in the fleet certainly represent less than 10% of all cars of this type.

   The mean CO level was $\mu = 2.9$ gm/mi, with standard deviation $\sigma = 0.4$ gm/mi. Since the sample size is large, the CLT allows us to model the sampling distribution of the $\bar{y}$ with a Normal model, with $\mu_{\bar{y}} = 2.9$ gm/mi and standard deviation $\sigma(\bar{y}) = \dfrac{0.4}{\sqrt{80}} = 0.045$ gm/mi.

b) According to the Normal model, the probability that $\bar{y}$ is between 3.0 and 3.1 gm/mi is approximately 0.0131.

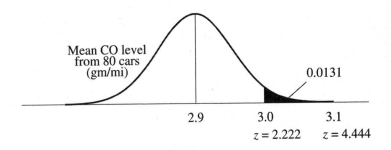

c) According to the Normal model, there is only a 5% chance that the fleet's mean CO level is greater than approximately 2.97 gm/mi.

$$z = \frac{\bar{y} - \mu_{\bar{y}}}{\sigma(\bar{y})}$$

$$1.645 = \frac{\bar{y} - 2.9}{0.045}$$

$$\bar{y} \approx 2.97$$

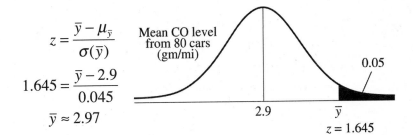

29. Tips.

a) Since the distribution of tips is skewed to the right, we can't use the Normal model to determine the probability that a given party will tip at least $20.

b) No. A sample of 4 parties is probably not a large enough sample for the CLT to allow us to use the Normal model to estimate the distribution of averages.

c) A sample of 10 parties may not be large enough to allow the use of a Normal model to describe the distribution of averages. It would be risky to attempt to estimate the probability that his next 10 parties tip an average of $15. However, since the distribution of tips has $\mu = \$9.60$, with standard deviation $\sigma = \$5.40$, we still know that the mean of the sampling distribution model is $\mu_{\bar{y}} = \$9.60$ with standard deviation $\sigma(\bar{y}) = \dfrac{5.40}{\sqrt{10}} \approx \$1.71$.

   We don't know the exact shape of the distribution, but we can still assess the likelihood of specific means. A mean tip of $15 is over 3 standard deviations above the expected mean tip for 10 parties. That's not very likely to happen.

31. More tips.

a) Random sampling condition: Assume that the tips from 40 parties can be considered a representative sample of all tips.

Independence assumption: It is reasonable to think that the tips are mutually independent, unless the service is particularly good or bad during this weekend.

10% condition: The tips of 40 parties certainly represent less than 10% of all tips.

The mean tip is $\mu = \$9.60$, with standard deviation $\sigma = \$5.40$. Since the sample size is large, the CLT allows us to model the sampling distribution of $\bar{y}$ with a Normal model, with $\mu_{\bar{y}} = \$9.60$ and standard deviation $\sigma(\bar{y}) = \dfrac{5.40}{\sqrt{40}} \approx \$0.8538$.

In order to earn at least $500, the waiter would have to average $\dfrac{500}{40} = \$12.50$ per party.

According to the Normal model, the probability that the waiter earns at least $500 in tips in a weekend is approximately 0.0003.

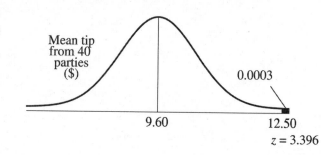

b) According to the Normal model, the waiter can expect to have a mean tip of about $10.6942, which corresponds to about $427.77 for 40 parties, in the best 10% of such weekends.

$$z = \frac{\bar{y} - \mu_{\bar{y}}}{\sigma(\bar{y})}$$

$$1.2816 = \frac{\bar{y} - 9.60}{0.8538}$$

$$\bar{y} \approx 10.6942$$

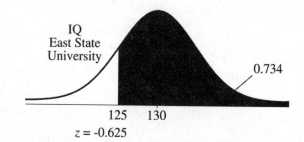

33. IQs.

a) According to the Normal model, the probability that the IQ of a student from East State is at least 125 is approximately 0.734.

b) First, we will need to generate a model for the difference in IQ between the two schools. Since we are choosing at random, it is reasonable to believe that the students' IQs are independent, which allows us to calculate the standard deviation of the difference.

$$\mu = E(E - W) = E(E) - E(W) = 130 - 120 = 10$$

$$\sigma = SD(E - W) = \sqrt{Var(E) + Var(W)}$$

$$= \sqrt{8^2 + 10^2} \approx 12.806$$

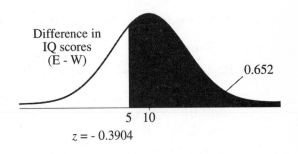

Since both distributions are Normal, the distribution of the difference is $N(10, 12.806)$.

According to the Normal model, the probability that the IQ of a student at ESU is at least 5 points higher than a student at WSU is approximately 0.652.

c) **Random sampling condition:** Students are randomly sampled from WSU.

Independence assumption: It is reasonable to think that the IQs are mutually independent.

10% condition: The 3 students certainly represent less than 10% of students.

The mean IQ is $\mu_w = 120$, with standard deviation $\sigma_w = 10$. Since the distribution IQs is Normal, we can model the sampling distribution of $\overline{w}$ with a Normal model, with $\mu_{\overline{w}} = 120$ with standard deviation

$$\sigma(\overline{w}) = \frac{10}{\sqrt{3}} \approx 5.7735.$$

According to the Normal model, the probability that the mean IQ of the 3 WSU students is above 125 is approximately 0.193.

d) As in part c, the sampling distribution of $\overline{e}$, the mean IQ of 3 ESU students, can be modeled with a Normal model, with $\mu_{\overline{e}} = 130$ with

standard deviation $\sigma(\overline{e}) = \dfrac{8}{\sqrt{3}} \approx 4.6188$.

The distribution of the difference in mean IQ is Normal, with the following parameters:

$$\mu_{\overline{e}-\overline{w}} = E(\overline{e} - \overline{w}) = E(\overline{e}) - E(\overline{w}) = 130 - 120 = 10$$

$$\sigma_{\overline{e}-\overline{w}} = SD(\overline{e} - \overline{w}) = \sqrt{Var(\overline{e}) + Var(\overline{w})}$$

$$= \sqrt{\left(\frac{10}{\sqrt{3}}\right)^2 + \left(\frac{8}{\sqrt{3}}\right)^2} \approx 7.3937$$

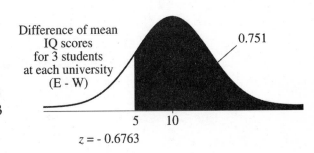

According to the Normal model, the probability that the mean IQ of 3 ESU students is at least 5 points higher than the mean IQ of 3 WSU students is approximately 0.751.

Chapter 19 – Confidence Intervals for Proportions

1. Margin of error.

He believes the true proportion of voters with a certain opinion is within 4% of his estimate, with some degree of confidence, perhaps 95% confidence.

3. Conditions.

a) *Population* – all cars; *sample* – 134 cars actually stopped at the checkpoint; p – proportion of all cars with safety problems; $\hat{p}$ – proportion of cars in the sample that actually have safety problems (10.4%).
Plausible independence condition: There is no reason to believe that the safety problems of cars are related to each other.
Randomization condition: This sample is not random, so hopefully the cars stopped are representative of cars in the area.
10% condition: The 134 cars stopped represent a small fraction of all cars, certainly less than 10%.
Success/Failure condition: $n\hat{p} = 14$ and $n\hat{q} = 120$ are both greater than 10, so the sample is large enough.
A one-proportion z-interval can be created for the proportion of all cars in the area with safety problems.

b) *Population* – the general public; *sample* – 602 viewers that logged on to the Web site; p – proportion of the general public that support prayer in school; $\hat{p}$ - proportion of viewers that logged on to the Web site and voted that support prayer in schools (81.1%).
Randomization condition: This sample is not random, but biased by voluntary response. It would be very unwise to attempt to use this sample to infer anything about the opinion of the general public related to school prayer.

c) *Population* – parents at the school; *sample* – 380 parents who returned surveys; p – proportion of all parents in favor of uniforms; $\hat{p}$ – proportion of those who responded that are in favor of uniforms (60%).
Randomization condition: This sample is not random, but rather biased by nonresponse. There may be lurking variables that affect the opinions of parents who return surveys (and the children who deliver them!).
It would be very unwise to attempt to use this sample to infer anything about the opinion of the parents about uniforms.

d) *Population* – all freshmen enrollees at the college (not just one year); *sample* – 1632 freshmen during the specified year; p – proportion of all students who will graduate on time; $\hat{p}$ – proportion of students from that year who graduate on time (85.05%).
Plausible independence condition: It is reasonable to think that the abilities of students to graduate on time are mutually independent.
Randomization condition: This sample is not random, but this year's freshmen class is probably representative of freshman classes in other years.
10% condition: The 1632 students in that years freshmen class represent less than 10% of all possible students.

Success/Failure condition: $n\hat{p} = 1388$ and $n\hat{q} = 244$ are both greater than 10, so the sample is large enough.

A one-proportion z-interval can be created for the proportion of freshmen that graduate on time from this college.

5. **Conclusions.**

a) Not correct. This statement implies certainty. There is no level of confidence in the statement.

b) Not correct. Different samples will give different results. Many fewer than 95% of samples are expected to have *exactly* 88% on-time orders.

c) Not correct. A confidence interval should say something about the unknown population proportion, not the sample proportion in different samples.

d) Not correct. We *know* that 88% of the orders arrived on time. There is no need to make an interval for the sample proportion.

e) Not correct. The interval should be about the proportion of on-time orders, not the days.

7. **Confidence intervals.**

a) False. For a given sample size, higher confidence means a *larger* margin of error.

b) True. Larger samples lead to smaller standard errors, which lead to smaller margins of error.

c) True. Larger samples are less variable, which makes us more confident that a given confidence interval succeeds in catching the population proportion.

d) False. The margin of error decreases as the square root of the sample size increases. Halving the margin of error requires a sample four times as large as the original.

9. **Cars.**

We are 90% confident that between 29.9% and 47.0% of cars are made in Japan.

11. **Ghosts.**

a) $ME = z^* \times SE(\hat{p}) = z^* \times \sqrt{\dfrac{\hat{p}\hat{q}}{n}} = 1.645 \times \sqrt{\dfrac{(0.38)(0.62)}{1012}} \approx 2.5\%$

b) The pollsters are 90% confident that the true proportion of adults who believe in ghosts is within 2.5% of the estimated 38%.

c) A 99% confidence interval requires a larger margin of error. In order to increase confidence, the interval must be wider.

d) $ME = z^* \times SE(\hat{p}) = z^* \times \sqrt{\dfrac{\hat{p}\hat{q}}{n}} = 2.576 \times \sqrt{\dfrac{(0.38)(0.62)}{1012}} \approx 3.9\%$

e) Smaller margins of error will give us less confidence in the interval.

13. Teenage drivers.

a) **Plausible independence condition:** There is no reason to believe that accidents selected at random would be related to one another.
Randomization condition: The insurance company randomly selected 582 accidents.
10% condition: 582 accidents represent less than 10% of all accidents.
Success/Failure condition: $n\hat{p} = 91$ and $n\hat{q} = 491$ are both greater than 10, so the sample is large enough.

Since the conditions are met, we can use a one-proportion z-interval to estimate the percentage of accidents involving teenagers.

$$\hat{p} \pm z^* \sqrt{\frac{\hat{p}\hat{q}}{n}} = \left(\frac{91}{582}\right) \pm 1.960 \sqrt{\frac{\left(\frac{91}{582}\right)\left(\frac{491}{582}\right)}{582}} = (12.7\%, 18.6\%)$$

b) We are 95% confident that between 12.7% and 18.6% of all accidents involve teenagers.

c) About 95% of random samples of size 582 will produce intervals that contain the true proportion of accidents involving teenagers.

d) Our confidence interval contradicts the assertion of the politician. The figure quoted by the politician, 1 out of every 5, or 20%, is outside the interval.

15. Safe food.

The grocer can conclude nothing about the opinions of all his customers from this survey. Those customers who bothered to fill out the survey represent a voluntary response sample, consisting of people who felt strongly one way or another about irradiated food. The randomization condition was not met.

17. Marijuana, again.

a) **Plausible independence condition:** There is no reason to believe that one randomly selected person's response will affect another's.
Randomization condition: The pollsters randomly selected 505 respondents.
10% condition: 505 respondents is less than 10% of all adults.
Success/Failure condition: $n\hat{p} = 505(0.31) = 157$ and $n\hat{q} = 505(0.69) = 348$ are both greater than 10, so the sample is large enough.

Since the conditions are met, we can use a one-proportion z-interval to estimate the percentage of people who think "the use of marijuana should be made legal."

$$\hat{p} \pm z^* \sqrt{\frac{\hat{p}\hat{q}}{n}} = (0.31) \pm 1.960 \sqrt{\frac{(0.31)(0.69)}{505}} = (27.0\%, 35.0\%)$$

We are 95% confident that between 27% and 35% of people think "the use of marijuana should be made legal."

b) The percentage of 47% does not fall within the interval created in part a.

19. Marijuana poll.

a) There may be response bias based on the wording of the question.

b) $\hat{p} \pm z^* \sqrt{\dfrac{\hat{p}\hat{q}}{n}} = (0.39) \pm 1.960 \sqrt{\dfrac{(0.39)(0.61)}{1010}} = (36.0\%, 42.0\%)$

c) The margin of error based on the pooled sample is smaller, since the sample size is larger.

21. Rickets.

a) **Plausible independence condition:** It is reasonable to think that the randomly selected children are mutually independent in regards to vitamin D deficiency.
Randomization condition: The 2,700 children were chosen at random.
10% condition: 2,700 children are less than 10% of all English children.
Success/Failure condition: $n\hat{p} = (2,700)(0.20) = 540$ and $n\hat{q} = (2,700)(0.80) = 2160$ are both greater than 10, so the sample is large enough.

Since the conditions are met, we can use a one-proportion z-interval to estimate the proportion of the English children with vitamin D deficiency.

$\hat{p} \pm z^* \sqrt{\dfrac{\hat{p}\hat{q}}{n}} = (0.20) \pm 2.326 \sqrt{\dfrac{(0.20)(0.80)}{2700}} = (18.2\%, 21.8\%)$

b) We are 98% confident that between 18.2% and 21.8% of English children are deficient in vitamin D.

c) About 98% of random samples of size 2,700 will produce confidence intervals that contain the true proportion of English children that are deficient in vitamin D.

23. Deer ticks.

a) **Plausible independence condition:** Deer ticks are parasites. A deer carrying the parasite may spread it to others. Ticks may not be distributed evenly throughout the population.
Randomization condition: The sample is not random and may not represent all deer.
10% condition: 153 deer are less than 10% of all deer.
Success/Failure condition: $n\hat{p} = 32$ and $n\hat{q} = 121$ are both greater than 10, so the sample is large enough.

The conditions are not satisfied, so we should use caution when a one-proportion z-interval is used to estimate the proportion of deer carrying ticks.

$\hat{p} \pm z^* \sqrt{\dfrac{\hat{p}\hat{q}}{n}} = \left(\dfrac{32}{153}\right) \pm 1.645 \sqrt{\dfrac{\left(\frac{32}{153}\right)\left(\frac{121}{153}\right)}{153}} = (15.5\%, 26.3\%)$

We are 90% confident that between 15.5% and 26.3% of deer have ticks.

b) In order to cut the margin of error in half, they must sample 4 times as many deer. $4(153) = 612$ deer.

c) The incidence of deer ticks is not plausibly independent, and the sample may not be representative of all deer, since females and young deer are usually not hunted.

25. Graduation.

a)

$$ME = z^* \sqrt{\frac{\hat{p}\hat{q}}{n}}$$

$$0.06 = 1.645 \sqrt{\frac{(0.25)(0.75)}{n}}$$

$$n = \frac{(1.645)^2 (0.25)(0.75)}{(0.06)^2}$$

$$n \approx 141 \text{ people}$$

In order to estimate the proportion of non-graduates in the 25-to 30-year-old age group to within 6% with 90% confidence, we would need a sample of at least 141 people. All decimals in the final answer must be rounded up, to the next person.
(For a more cautious answer, let $\hat{p} = \hat{q} = 0.5$. This method results in a required sample of 188 people.)

b)

$$ME = z^* \sqrt{\frac{\hat{p}\hat{q}}{n}}$$

$$0.04 = 1.645 \sqrt{\frac{(0.25)(0.75)}{n}}$$

$$n = \frac{(1.645)^2 (0.25)(0.75)}{(0.04)^2}$$

$$n \approx 318 \text{ people}$$

In order to estimate the proportion of non-graduates in the 25-to 30-year-old age group to within 4% with 90% confidence, we would need a sample of at least 318 people. All decimals in the final answer must be rounded up, to the next person.
(For a more cautious answer, let $\hat{p} = \hat{q} = 0.5$. This method results in a required sample of 423 people.)
Alternatively, the margin of error is now 2/3 of the original, so the sample size must be increased by a factor of 9/4. $141(9/4) \approx 318$ people.

c)

$$ME = z^* \sqrt{\frac{\hat{p}\hat{q}}{n}}$$

$$0.03 = 1.645 \sqrt{\frac{(0.25)(0.75)}{n}}$$

$$n = \frac{(1.645)^2 (0.25)(0.75)}{(0.03)^2}$$

$$n \approx 564 \text{ people}$$

In order estimate the proportion of non-graduates in the 25-to 30-year-old age group to within 3% with 90% confidence, we would need a sample of at least 564 people. All decimals in the final answer must be rounded up, to the next person.
(For a more cautious answer, let $\hat{p} = \hat{q} = 0.5$. This method results in a required sample of 752 people.)

Alternatively, the margin of error is now half that of the original, so the sample size must be increased by a factor of 4. $141(4) \approx 564$ people.

27. Graduation, again.

$$ME = z^* \sqrt{\frac{\hat{p}\hat{q}}{n}}$$

$$0.02 = 1.960 \sqrt{\frac{(0.25)(0.75)}{n}}$$

$$n = \frac{(1.960)^2 (0.25)(0.75)}{(0.02)^2}$$

$$n \approx 1,801 \text{ people}$$

In order to estimate the proportion of non-graduates in the 25-to 30-year-old age group to within 2% with 95% confidence, we would need a sample of at least 1,801 people. All decimals in the final answer must be rounded up, to the next person.
(For a more cautious answer, let $\hat{p} = \hat{q} = 0.5$. This method results in a required sample of 2,401 people.)

29. Pilot study.

$$ME = z^*\sqrt{\frac{\hat{p}\hat{q}}{n}}$$

$$0.03 = 1.645\sqrt{\frac{(0.15)(0.85)}{n}}$$

$$n = \frac{(1.645)^2(0.15)(0.85)}{(0.03)^2}$$

$$n \approx 384 \text{ cars}$$

Use $\hat{p} = \frac{9}{60} = 0.15$ from the pilot study as an estimate.

In order to estimate the percentage of cars with faulty emissions systems to within 3% with 90% confidence, the state's environmental agency will need a sample of at least 384 cars. All decimals in the final answer must be rounded up, to the next car.

31. Approval rating.

$$ME = z^*\sqrt{\frac{\hat{p}\hat{q}}{n}}$$

$$0.025 = z^*\sqrt{\frac{(0.65)(0.35)}{972}}$$

$$z^* = \frac{0.025}{\sqrt{\frac{(0.65)(0.35)}{972}}}$$

$$z^* \approx 1.634$$

Since $z^* \approx 1.634$, which is close to 1.645, the pollsters were probably using 90% confidence. The slight difference in the z^* values is due to rounding of the governor's approval rating.

Chapter 20 – Testing Hypotheses about Proportions

1. **Hypotheses.**

   a) H_0 : The governor's "negatives" are 30%. ($p = 0.30$)
   H_A : The governor's "negatives" are less than 30%. ($p < 0.30$)

   b) H_0 : The proportion of heads is 50%. ($p = 0.50$)
   H_A : The proportion of heads is not 50%. ($p \neq 0.50$)

   c) H_0 : The proportion of people who quit smoking is 20%. ($p = 0.20$)
   H_A : The proportion of people who quit smoking is greater than 20%. ($p > 0.20$)

3. **Negatives.**

   Statement d is the correct interpretation of a *P*-value.

5. **Relief.**

   It is *not* reasonable to conclude that the new formula and the old one are equally effective. Furthermore, our inability to make that conclusion has nothing to do with the *P*-value. We can not prove the null hypothesis (that the new formula and the old formula are equally effective), but can only fail to find evidence that would cause us to reject it. All we can say about this *P*-value is that there is a 27% chance of seeing the observed effectiveness from natural sampling variation if the new formula and the old one are equally effective.

7. **Blinking timers.**

   1) Null and alternative hypotheses should involve p, not $\hat{p}$.

   2) The company wants *at least* 90% of its customers to succeed. H_A should be $p > 0.90$.

   3) The student failed to check $nq = (200)(0.10) = 20 > 10$.

   4) $SD(\hat{p}) = \sqrt{\dfrac{pq}{n}} = \sqrt{\dfrac{(0.90)(0.10)}{200}} \approx 0.021$. The student used $\hat{p}$ and $\hat{q}$.

   5) Value of z is incorrect. The correct value is $z = \dfrac{0.94 - 0.90}{0.021} \approx 1.90$.

   6) *P*-value is incorrect. $P = P(z > 1.90) = 0.029$

   7) For the *P*-value given, an incorrect conclusion is drawn. A *P*-value of 0.009 provides evidence that the new system works. We can never make a conclusion like the one given, regardless of the *P*-value. The most we can ever say (for high *P*-values) is that there is no evidence that the new system works. The correct conclusion for the corrected *P*-value is: Since the *P*-value is low, there is strong evidence that the new system works.

9. Dowsing.

a) H_0 : The percentage of successful wells drilled by the dowser is 30%. ($p = 0.30$)
H_A : The percentage of successful wells drilled by the dowser is greater than 30%. ($p > 0.30$)

b) **Plausible independence condition:** There is no reason to think that finding water in one well will affect the probability that water is found in another, unless the wells are close enough to be fed by the same underground water source.
Randomization condition: This sample is not random, so hopefully the customers you check with are representative of all of the dowser's customers.
10% condition: The 80 customers sampled are less than 10% of all possible customers.
Success/Failure condition: $np = (80)(0.30) = 24$ and $nq = (80)(0.70) = 56$ are both greater than 10, so the sample is large enough.

c) The sample of customers may not be representative of all customers, so we will proceed cautiously. A Normal model can be used to model the sampling distribution of the proportion, with $\mu_{\hat{p}} = p = 0.30$ and $\sigma(\hat{p}) = \sqrt{\dfrac{pq}{n}} = \sqrt{\dfrac{(0.30)(0.70)}{80}} \approx 0.0512$.

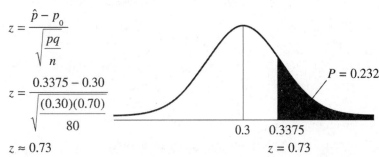

We can perform a one-proportion z-test. The observed proportion of successful wells is

$\hat{p} = \dfrac{27}{80} = 0.3375$.

$z = \dfrac{\hat{p} - p_0}{\sqrt{\dfrac{pq}{n}}}$

$z = \dfrac{0.3375 - 0.30}{\sqrt{\dfrac{(0.30)(0.70)}{80}}}$

$z \approx 0.73$

d) If his dowsing has the same success rate as standard drilling methods, there is more than a 23% chance of seeing results as good as those of the dowser, or better, by natural sampling variation.

e) With a *P*-value of 0.232, we fail to reject the null hypothesis. There is no evidence to suggest that the dowser has a success rate any higher than 30%.

11. Smoking.

a) **Plausible independence condition:** There is no reason to believe that one randomly selected adult's response will affect another's, with regards to smoking.
Randomization condition: The health survey used 881 randomly selected adults.
10% condition: 881 adults is less than 10% of all adults.
Success/Failure condition: $n\hat{p} = 458$ and $n\hat{q} = 423$ are both greater than 10, so the sample is large enough.

Since the conditions are met, we can use a one-proportion *z*-interval to estimate the proportion of the adults who have never smoked.

$$\hat{p} \pm z^{*} \sqrt{\dfrac{\hat{p}\hat{q}}{n}} = (0.52) \pm 1.960 \sqrt{\dfrac{(0.52)(0.48)}{881}} = (48.7\%, 55.3\%)$$

We are 95% confident that between 48.7% and 55.3% of adults have never smoked.

b) H_0 : The percentage of adults who smoke is 44%. ($p = 0.44$)
H_A : The percentage of adults who smoke is not 44%. ($p \neq 0.44$)

Since 44% is not in the 95% confidence interval, we will reject the null hypothesis. There is strong evidence that, in 1995, the percentage of adults who have never smoked was not 44%. In fact, our sample indicates an increase (since the 1960s) in the percentage of adults who have never smoked.

13. Pollution.

H_0 : The percentage of cars with faulty emissions is 20%. ($p = 0.20$)
H_A : The percentage of cars with faulty emissions is greater than 20%. ($p > 0.20$)

Two conditions are not satisfied. 22 is greater than 10% of the population of 150 cars, and $np = (22)(0.20) = 4.4$, which is not greater than 10. It's probably not a good idea to proceed with a hypothesis test.

15. Twins.

H_0 : The percentage of twin births to teenage girls is 3%. ($p = 0.03$)
H_A : The percentage of twin births to teenage girls is less than 3%. ($p < 0.03$)

Plausible independence condition: One mother having twins will not affect another. Observations are plausibly independent.
Randomization condition: This sample may not be random, but it is reasonable to think that this hospital has a representative sample of teen mothers, with regards to twin births.
10% condition: The sample of 469 teenage mothers is less than 10% of all such mothers.
Success/Failure condition: $np = (469)(0.03) = 14.07$ and $nq = (469)(0.97) = 454.93$ are both greater than 10, so the sample is large enough.

The conditions have been satisfied, so a Normal model can be used to model the sampling distribution of the proportion, with $\mu_{\hat{p}} = p = 0.03$ and $\sigma(\hat{p}) = \sqrt{\dfrac{pq}{n}} = \sqrt{\dfrac{(0.03)(0.97)}{469}} \approx 0.0079$.

We can perform a one-proportion z-test. The observed proportion of twin births to teenage mothers is $\hat{p} = \dfrac{7}{469} \approx 0.015$.

Since the *P*-value = 0.0278 is low, we reject the null hypothesis. There is evidence that the proportion of twin births for teenage mothers at this large city hospital is lower than the proportion of twin births for all mothers.

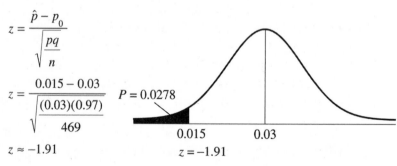

$$z = \frac{\hat{p} - p_0}{\sqrt{\dfrac{pq}{n}}}$$

$$z = \frac{0.015 - 0.03}{\sqrt{\dfrac{(0.03)(0.97)}{469}}}$$

$$z \approx -1.91$$

$P = 0.0278$

0.015   0.03
$z = -1.91$

17. Webzine.

H_0 : The percentage of readers interested in an online edition is 25%. ($p = 0.25$)
H_A : The percentage of readers interested in an online edition is greater than 25%. ($p > 0.25$)

Plausible independence condition: Interest of one reader should not affect interest of other readers.
Randomization condition: The magazine conducted an SRS of 500 current readers.
10% condition: 500 readers are less than 10% of all potential subscribers.
Success/Failure condition: $np = (500)(0.25) = 125$ and $nq = (500)(0.75) = 375$ are both greater than 10, so the sample is large enough.

The conditions have been satisfied, so a Normal model can be used to model the sampling distribution of the proportion, with $\mu_{\hat{p}} = p = 0.25$ and $\sigma(\hat{p}) = \sqrt{\dfrac{pq}{n}} = \sqrt{\dfrac{(0.25)(0.75)}{500}} \approx 0.0194$.

We can perform a one-proportion z-test. The observed proportion of interested readers is $\hat{p} = \dfrac{137}{500} = 0.274$.

Since the *P*-value = 0.1076 is high, we fail to reject the null hypothesis. There is little evidence to suggest that the proportion of interested readers is greater than 25%. The magazine should not publish the online edition.

$$z = \dfrac{\hat{p} - p_0}{\sqrt{\dfrac{pq}{n}}}$$

$$z = \dfrac{0.274 - 0.25}{\sqrt{\dfrac{(0.25)(0.75)}{500}}}$$

$$z \approx 1.24$$

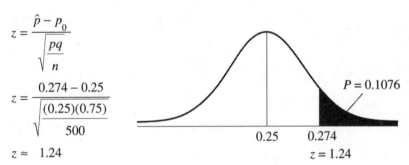

$P = 0.1076$

0.25   0.274
$z = 1.24$

19. Women executives.

H_0 : The proportion of female executives is similar to the overall proportion of female employees at the company. ($p = 0.40$)
H_A : The proportion of female executives is lower than the overall proportion of female employees at the company. ($p < 0.40$)

Plausible independence condition: It is reasonable to think that executives at this company were chosen independently.
Randomization condition: The executives were not chosen randomly, but it is reasonable to think of these executives as representative of all potential executives over many years.
10% condition: 43 executives are less than 10% of all possible executives at the company.
Success/Failure condition: $np = (43)(0.40) = 17.2$ and $nq = (43)(0.60) = 25.8$ are both greater than 10, so the sample is large enough.

The conditions have been satisfied, so a Normal model can be used to model the sampling distribution of the proportion, with $\mu_{\hat{p}} = p = 0.40$ and $\sigma(\hat{p}) = \sqrt{\dfrac{pq}{n}} = \sqrt{\dfrac{(0.40)(0.60)}{43}} \approx 0.0747$.

We can perform a one-proportion z-test. The observed proportion of female executives is $\hat{p} = \dfrac{13}{43} \approx 0.302$.

Since the *P*-value = 0.0955 is high, we fail to reject the null hypothesis. There is little evidence to suggest proportion of female executives is any different from the overall proportion of 40% female employees at the company.

$$z = \frac{\hat{p} - p_0}{\sqrt{\dfrac{pq}{n}}}$$

$$z = \frac{0.302 - 0.40}{\sqrt{\dfrac{(0.40)(0.60)}{43}}}$$

$$z \approx -1.31$$

$P = 0.0955$

$z = -1.31$

21. Dropouts.

H_0 : The proportion of dropouts at this high school is similar to 10.9%, the proportion of dropouts nationally. ($p = 0.109$)

H_A : The proportion of dropouts at this high school is greater than 10.9%, the proportion of dropouts nationally. ($p > 0.109$)

Plausible independence condition/Randomization condition: Assume that the students at this high school are representative of all students nationally. This is really what we are testing. The dropout rate at this high school has traditionally been close to the national rate. If we reject the null hypothesis, we will have evidence that the dropout rate at this high school is no longer close to the national rate.

10% condition: 1792 students are less than 10% of all students nationally.

Success/Failure condition: $np = (1782)(0.109) = 194.238$ and $nq = (1782)(0.891) = 1587.762$ are both greater than 10, so the sample is large enough.

The conditions have been satisfied, so a Normal model can be used to model the sampling distribution of the proportion, $\mu_{\hat{p}} = p = 0.109$ and $\sigma(\hat{p}) = \sqrt{\dfrac{pq}{n}} = \sqrt{\dfrac{(0.109)(0.891)}{1782}} \approx 0.0074$.

We can perform a one-proportion *z*-test. The observed proportion of dropouts is $\hat{p} = \dfrac{210}{1782} \approx 0.117845$.

Since the *P*-value = 0.115 is high, we fail to reject the null hypothesis. There is little evidence of an increase in dropout rate from 10.9%.

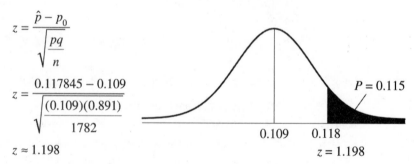

$$z = \frac{\hat{p} - p_0}{\sqrt{\dfrac{pq}{n}}}$$

$$z = \frac{0.117845 - 0.109}{\sqrt{\dfrac{(0.109)(0.891)}{1782}}}$$

$$z \approx 1.198$$

$P = 0.115$

$0.109 \quad 0.118$

$z = 1.198$

23. Lost luggage.

H_0 : The proportion of lost luggage returned the next day is 90%. ($p = 0.90$)
H_A : The proportion of lost luggage returned the next day is lower than 90%. ($p < 0.90$)

Plausible independence condition: It is reasonable to think that the people surveyed were independent with regards to their luggage woes.
Randomization condition: Although not stated, we will hope that the survey was conducted randomly, or at least that these air travelers are representative of all air travelers for that airline.
10% condition: 122 air travelers are less than 10% of all air travelers on the airline.
Success/Failure condition: $np = (122)(0.90) = 109.8$ and $nq = (122)(0.10) = 12.2$ are both greater than 10, so the sample is large enough.

The conditions have been satisfied, so a Normal model can be used to model the sampling distribution of the proportion, with $\mu_{\hat{p}} = p = 0.90$ and $\sigma(\hat{p}) = \sqrt{\dfrac{pq}{n}} = \sqrt{\dfrac{(0.90)(0.10)}{122}} \approx 0.0272$.

We can perform a one-proportion z-test. The observed proportion of dropouts is $\hat{p} = \dfrac{103}{122} \approx 0.844$.

Since the *P*-value = 0.0201 is low, we reject the null hypothesis. There is evidence that the proportion of lost luggage returned the next day is lower than the 90% claimed by the airline.

$$z = \frac{\hat{p} - p_0}{\sqrt{\dfrac{pq}{n}}}$$

$$z = \frac{0.844 - 0.90}{\sqrt{\dfrac{(0.90)(0.10)}{122}}}$$

$$z \approx -2.05$$

$P = 0.0201$

$z = -2.05$

Chapter 21 – More About Tests

1. **P-value.**

   If the effectiveness of the new poison ivy treatment is the same as the effectiveness of the old treatment, the chance of observing an effectiveness this large or larger in a sample of the same size is 4.7% by natural sampling variation alone.

3. **Alpha.**

   Since the null hypothesis was rejected at $\alpha = 0.05$, the P-value for the researcher's test must have been less than 0.05. He would have made the same decision at $\alpha = 0.10$, since the P-value must also be less than 0.10. We can't be certain whether or not he would have made the same decision at $\alpha = 0.01$, since we only know that the P-value was less than 0.05. It may have been less than 0.01, but we can't be sure.

5. **Significant?**

   a) If 90% of children have really been vaccinated, there is only a 1.1% chance of observing 89.4% of children (in a sample of 13,000) vaccinated by natural sampling variation alone.

   b) We conclude that the proportion of children who have been vaccinated is below 90%, but a 95% confidence interval would show that the true proportion is between 88.9% and 89.9%. Most likely a decrease from 90% to 89.9% would not be considered important. The 90% figure was probably an approximate figure anyway.

7. **Testing cars.**

   H_0 : The shop is meeting the emissions standards.
   H_A : The shop is not meeting the emissions standards.

   a) Type I error is when the regulators decide that the shop is not meeting standards when they actually are meeting the standards.

   b) Type II error is when the regulators certify the shop when they are not meeting the standards.

   c) Type I would be more serious to the shop owners. They would lose their certification, even though they are meeting the standards.

   d) Type II would be more serious to environmentalists. Shops are allowed to operate, even though they are allowing polluting cars to operate.

9. **Cars again.**

   a) The power of the test is the probability of detecting that the shop is not meeting standards when they are not.

   b) The power of the test will be greater when 40 cars are tested. A larger sample size increases the power of the test.

   c) The power of the test will be greater when the level of significance is 10%. There is a greater chance that the null hypothesis will be rejected.

   d) The power of the test will be greater when the shop is out of compliance "a lot". Larger problems are easier to detect.

11. Equal opportunity?

H_0 : The company is not discriminating against minorities.
H_A : The company is discriminating against minorities.

a) This is a one-tailed test. They wouldn't sue if "too many" minorities were hired.

b) Type I error would be deciding that the company is discriminating against minorities when they are not discriminating.

c) Type II error would be deciding that the company is not discriminating against minorities when they actually are discriminating.

d) The power of the test is the probability that discrimination is detected when it is actually occurring.

e) The power of the test will increase when the level of significance is increased from 0.01 to 0.05.

f) The power of the test is lower when the lawsuit is based on 37 employees instead of 87. Lower sample size leads to less power.

13. Dropouts.

a) The test is one-tailed. We are testing to see if a decrease in the dropout rate is associated with the software.

b) H_0 : The dropout rate does not change following the use of the software. ($p = 0.13$)
H_A : The dropout rate decreases following the use of the software. ($p < 0.13$)

c) The professor makes a Type I error if he buys the software when the dropout rate has not actually decreased.

d) The professor makes a Type II error if he doesn't buy the software when the dropout rate has actually decreased.

e) The power of the test is the probability of buying the software when the dropout rate has actually decreased.

15. Dropouts, part II.

a) H_0 : The dropout rate does not change following the use of the software. ($p = 0.13$)
H_A : The dropout rate decreases following the use of the software. ($p < 0.13$)

   Plausible independence assumption: One student's decision about dropping out should not influence another's decision.
   Randomization condition: This year's class of 203 students is probably representative of all stats students.
   10% condition: A sample of 203 students is less than 10% of all students.
   Success/Failure condition: $np = (203)(0.13) = 26.39$ and $nq = (203)(0.87) = 176.61$ are both greater than 10, so the sample is large enough.

The conditions have been satisfied, so a Normal model can be used to model the sampling distribution of the proportion, with $\mu_{\hat{p}} = p = 0.13$ and $\sigma(\hat{p}) = \sqrt{\dfrac{pq}{n}} = \sqrt{\dfrac{(0.13)(0.87)}{203}} \approx 0.0236$.

We can perform a one-proportion z-test. The observed proportion of dropouts is $\hat{p} = \dfrac{11}{203} \approx 0.054$.

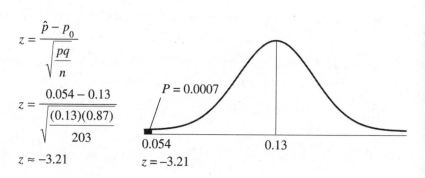

Since the P-value = 0.0007 is very low, we reject the null hypothesis. There is strong evidence that the dropout rate has dropped since use of the software program was implemented. As long as the professor feels confident that this class of stats students is representative of all potential students, then he should buy the program.

$$z = \dfrac{\hat{p} - p_0}{\sqrt{\dfrac{pq}{n}}}$$

$$z = \dfrac{0.054 - 0.13}{\sqrt{\dfrac{(0.13)(0.87)}{203}}}$$

$$z \approx -3.21$$

$P = 0.0007$

$z = -3.21$

If you used a 95% confidence interval to assess the effectiveness of the program:

$$\hat{p} \pm z^* \sqrt{\dfrac{\hat{p}\hat{q}}{n}} = \left(\dfrac{11}{203}\right) \pm 1.960 \sqrt{\dfrac{\left(\frac{11}{203}\right)\left(\frac{192}{203}\right)}{203}} = (2.3\%, 8.5\%)$$

We are 95% confident that the dropout rate is between 2.3% and 8.5%. Since 15% is not contained in the interval, this provides evidence that the dropout rate has changed following the implementation of the software program.

b) The chance of observing 11 or fewer dropouts in a class of 203 is only 0.07% if the dropout rate is really 13%.

17. Hoops.

H_0 : The player's foul-shot percentage is only 60%. ($p = 0.60$)
H_A : The player's foul-shot percentage is better than 60%. ($p > 0.60$)

a) The player's shots can be considered Bernoulli trials. There are only two possible outcomes, make the shot and miss the shot. The probability of making any shot is constant at $p = 0.60$. Assume that the shots are independent of each other. Use $Binom(10, 0.60)$.

Let X = the number of shots made out of $n = 10$.

$$P(\text{makes at least 9 out of 10}) = P(X \geq 9)$$
$$= P(X = 9) + P(X = 10)$$
$$= {}_{10}C_9(0.60)^9(0.40)^1 + {}_{10}C_{10}(0.60)^{10}(0.40)^0$$
$$\approx 0.0464$$

b) The coach made a Type I error.

c) The power of the test can be calculated for specific values of the new probability of success. Each true value of p has a power calculation associated with it. In this case, we are finding the power of the test to detect an 80% foul-shooter. Use $Binom(10, 0.80)$.

Let X = the number of shots made out of $n = 10$.

$$P(\text{makes at least 9 out of 10}) = P(X \geq 9)$$
$$= P(X = 9) + P(X = 10)$$
$$= {}_{10}C_9(0.80)^9(0.20)^1 + {}_{10}C_{10}(0.80)^{10}(0.20)^0$$
$$\approx 0.376$$

The power of the test to detect an increase in foul-shot percentage from 60% to 80% is about 37.6%.

d) The power of the test to detect improvement in foul-shooting can be increased by increasing the number of shots, or by keeping the number of shots at 10 but increasing the level of significance by declaring that 8, 9, or 10 shots made will convince the coach that the player has improved. In other words, the coach can increase the power of the test by lowering the standard of proof.

19. Survey.

Only the Laborers data may be used to construct a confidence interval.

The Laborers data are okay, since $n = 300$ is less than 10% of 6235 employees, and $n\hat{p} = (300)(0.09) = 27$ and $nq = (300)(0.91) = 273$ are both greater than 10, so the sample is large enough. (We are assuming that the samples are random and free from biases.)

The Clerical data are not okay, since $n = 200$ is greater than 10% of 1520 employees.

The Management data are not okay, since $n\hat{p} = (25)(0.08) = 2$, which is not greater than 10. The Success/Failure condition is not satisfied.

21. Little league.

a) Plausible independence: It is reasonable to think that one pitcher's condition would not affect the conditions of other pitchers.
Randomization condition: The study is not identified at random. At the very least, let's hope that the sample is representative of all young pitchers.
10% condition: 298 young pitchers are less than 10% of all young pitchers.
Success/Failure condition: $n\hat{p} = (0.26)(298) = 77$ and $n\hat{q} = (0.74)(298) = 221$ are both greater than 10, so the sample is large enough.

Since the conditions are met, we can use a one-proportion z-interval to estimate the proportion of young pitchers who complained of elbow pain.

$$\hat{p} \pm z^* \sqrt{\frac{\hat{p}\hat{q}}{n}} = (0.26) \pm 1.645 \sqrt{\frac{(0.26)(0.74)}{298}} = (21.8\%, 30.2\%)$$

We are 90% confident that between 21.8% and 30.2% of young pitchers complained of elbow pain.

b) The coaches claim that "only" 1 kid in 5 is at risk of arm injury is not consistent with interval, since 20% is below the lower end of the interval.

Chapter 22 – Comparing Two Proportions

1. **Arthritis.**

   a) **Randomization condition:** Americans age 65 and older were selected randomly.
      10% condition: 1012 men and 1062 women are less than 10% of all men and women.
      Independent samples condition: The sample of men and the sample of women were drawn independently of each other.
      Success/Failure condition: $n\hat{p}$ (men) = 411, $n\hat{q}$ (men) = 601, $n\hat{p}$ (women) = 535, and $n\hat{q}$ (women) = 527 are all greater than 10, so the samples are both large enough.

      Since the conditions have been satisfied, we will find a two-proportion z-interval.

   b) $(\hat{p}_F - \hat{p}_M) \pm z^* \sqrt{\dfrac{\hat{p}_F \hat{q}_F}{n_F} + \dfrac{\hat{p}_M \hat{q}_M}{n_M}} = \left(\dfrac{535}{1062} - \dfrac{411}{1012}\right) \pm 1.960 \sqrt{\dfrac{\left(\frac{535}{1062}\right)\left(\frac{527}{1062}\right)}{1062} + \dfrac{\left(\frac{411}{1012}\right)\left(\frac{601}{1012}\right)}{1012}} = (0.055, 0.140)$

   c) We are 95% confident that the proportion of American women age 65 and older who suffer from arthritis is between 5.5% and 14.0% higher than the proportion of American men the same age who suffer from arthritis.

   d) Since the interval for the difference in proportions of arthritis sufferers does not contain 0, there is strong evidence that arthritis is more likely to afflict women than men.

3. **Pets.**

   a) $SE(\hat{p}_{Herb} - \hat{p}_{None}) = \sqrt{\dfrac{\hat{p}_{Herb} \hat{q}_{Herb}}{n_{Herb}} + \dfrac{\hat{p}_{None} \hat{q}_{None}}{n_{None}}} = \sqrt{\dfrac{\left(\frac{473}{827}\right)\left(\frac{354}{827}\right)}{827} + \dfrac{\left(\frac{19}{130}\right)\left(\frac{111}{130}\right)}{130}} = 0.035$

   b) **Randomization condition:** Assume that the dogs studied were representative of all dogs.
      10% condition: 827 dogs from homes with herbicide used regularly and 130 dogs from homes with no herbicide used are less than 10% of all dogs.
      Independent samples condition: The samples were drawn independently of each other.
      Success/Failure condition: $n\hat{p}$ (herb) = 473, $n\hat{q}$ (herb) = 354, $n\hat{p}$ (none) = 19, and $n\hat{q}$ (none) = 111 are all greater than 10, so the samples are both large enough.

      Since the conditions have been satisfied, we will find a two-proportion z-interval.

   $(\hat{p}_{Herb} - \hat{p}_{None}) \pm z^* \sqrt{\dfrac{\hat{p}_{Herb} \hat{q}_{Herb}}{n_{Herb}} + \dfrac{\hat{p}_{None} \hat{q}_{None}}{n_{None}}}$

   $= \left(\dfrac{473}{827} - \dfrac{19}{130}\right) \pm 1.960 \sqrt{\dfrac{\left(\frac{473}{827}\right)\left(\frac{354}{827}\right)}{827} + \dfrac{\left(\frac{19}{130}\right)\left(\frac{111}{130}\right)}{130}} = (0.356, 0.495)$

   c) We are 95% confident that the proportion of pets with a malignant lymphoma in homes where herbicides are used is between 35.6% and 49.5% higher than the proportion of pets with lymphoma in homes where no pesticides are used.

5. **Prostate cancer.**

   a) This was an experiment. Men were randomly assigned to imposed treatments. They were assigned to either have prostate surgery or assigned to not have prostate surgery.

   b) **Randomization condition:** The men were randomly assigned to the two treatment groups.
   10% condition: 347 men who had surgery and 348 men who did not have surgery are both less than 10% of all men.
   Independent samples condition: The groups were assigned randomly, so the groups are not related.
   Success/Failure condition: $n\hat{p}$ (surg) = 16, $n\hat{q}$ (surg) = 331, $n\hat{p}$ (none) = 31, and $n\hat{q}$ (splint) = 317 are all greater than 10, so the samples are both large enough.

   Since the conditions have been satisfied, we will find a two-proportion z-interval.

   $$\left(\hat{p}_{None} - \hat{p}_{Surg}\right) \pm z^* \sqrt{\frac{\hat{p}_{None}\hat{q}_{None}}{n_{None}} + \frac{\hat{p}_{Surg}\hat{q}_{Surg}}{n_{Surg}}}$$

   $$= \left(\tfrac{31}{348} - \tfrac{16}{347}\right) \pm 1.960 \sqrt{\frac{\left(\tfrac{31}{348}\right)\left(\tfrac{317}{348}\right)}{348} + \frac{\left(\tfrac{16}{347}\right)\left(\tfrac{331}{347}\right)}{347}} = (0.006, 0.080)$$

   We are 95% confident that the proportion of patients who die from prostate cancer after having no surgery is between 0.60% and 8.0% higher than the proportion of patients who die after having surgery.

   c) Since 0 is not contained in the interval, there is evidence that surgery may be effective in preventing death from prostate cancer.

7. **Politics.**

   a) The margin of error is larger for the difference in proportions of support because differences have larger standard errors than single samples.

   b) The confidence interval for difference in support was 2% ± 4%, or (–2%, 6%). Since the interval contains 0, there is no evidence of a difference in the proportion of support for antiterrorist legislation.

9. **Teen smoking, part I.**

   a) This is a prospective observational study.

   b) H_0: The proportion of teen smokers among the group whose parents disapprove of smoking is the same as the proportion of teen smokers among the group whose parents are lenient about smoking. $\left(p_{Dis} = p_{Len} \ \text{ or } \ p_{Dis} - p_{Len} = 0\right)$

   H_A: The proportion of teen smokers among the group whose parents disapprove of smoking is lower than the proportion of teen smokers among the group whose parents are lenient about smoking. $\left(p_{Dis} < p_{Len} \ \text{ or } \ p_{Dis} - p_{Len} < 0\right)$

   c) **Randomization condition:** Assume that the teens surveyed are representative of all teens.
   10% condition: 284 and 41 are both less than 10% of all teens.
   Independent samples condition: The groups were surveyed independently.
   Success/Failure condition: $n\hat{p}$ (disapprove) = 54, $n\hat{q}$ (disapprove) = 230, $n\hat{p}$ (lenient) = 11, and $n\hat{q}$ (lenient) = 30 are all greater than 10, so the samples are both large enough.

Since the conditions have been satisfied, we will model the sampling distribution of the difference in proportion with a Normal model with mean 0 and standard deviation

$$\text{estimated by } SE_{pooled}\left(\hat{p}_{Dis} - \hat{p}_{Len}\right) = \sqrt{\frac{\hat{p}_{pooled}\hat{q}_{pooled}}{n_{Dis}} + \frac{\hat{p}_{pooled}\hat{q}_{pooled}}{n_{Len}}} = \sqrt{\frac{\left(\frac{65}{325}\right)\left(\frac{260}{325}\right)}{284} + \frac{\left(\frac{65}{325}\right)\left(\frac{260}{325}\right)}{41}} = 0.0668.$$

d) The observed difference between the proportions is 0.190 − 0.268 = − 0.078.

Since the *P*-value = 0.1211 is high, we fail to reject the null hypothesis. There is little evidence to suggest that parental attitudes influence teens' decisions to smoke.

$$z = \frac{-0.078 - 0}{0.0668}$$

$$z \approx -1.17$$

$P = 0.1211$

$-0.078 \quad 0$

$z = -1.17$

e) If there is no difference in the proportions, there is about a 12% chance of seeing the observed difference or larger by natural sampling variation.

f) If teens' decisions about smoking *are* influenced, we have committed a Type II error.

11. Teen smoking, part II.

a) Since the conditions have already been satisfied in Exercise 9, we will find a two-proportion *z*-interval.

$$\left(\hat{p}_{Dis} - \hat{p}_{Len}\right) \pm z^* \sqrt{\frac{\hat{p}_{Dis}\hat{q}_{Dis}}{n_{Dis}} + \frac{\hat{p}_{Len}\hat{q}_{Len}}{n_{Len}}}$$

$$= \left(\frac{54}{284} - \frac{11}{41}\right) \pm 1.960 \sqrt{\frac{\left(\frac{54}{284}\right)\left(\frac{230}{284}\right)}{284} + \frac{\left(\frac{11}{41}\right)\left(\frac{30}{41}\right)}{41}} = (-0.065, 0.221)$$

b) We are 95% confident that the proportion of teens whose parents disapprove of smoking who will eventually smoke is between 6.5% less and 22.1% more than for teens with parents who are lenient about smoking.

c) We expect 95% of random samples of this size to produce intervals that contain the true difference between the proportions.

13. Pregnancy.

a) H_0: The proportion of live births is the same for women under the age of 38 as it is for women 38 or older. $\left(p_{<38} = p_{\geq38} \text{ or } p_{<38} - p_{\geq38} = 0\right)$

H_A: The proportion of live births is different for women under the age of 38 than for women 38 or older. $\left(p_{<38} \neq p_{\geq38} \text{ or } p_{<38} - p_{\geq38} \neq 0\right)$

Randomization condition: Assume that the women studied are representative of all women.
10% condition: 157 and 89 are both less than 10% of all women.
Independent samples condition: The groups are not associated.
Success/Failure condition: $n\hat{p}$ (under 38) = 42, $n\hat{q}$ (under 38) = 115, $n\hat{p}$ (38 and over) = 7, and $n\hat{q}$ (38 and over) = 82 are not all greater than 10, since the observed number of live births is only 7. However, if we check the pooled value, $n\hat{p}_{pooled}$ (38 and over) = (89)(0.1992) = 18. All of the samples are large enough.

Since the conditions have been satisfied, we will model the sampling distribution of the difference in proportion with a Normal model with mean 0 and standard deviation

estimated by $SE_{pooled}\left(\hat{p}_{<38} - \hat{p}_{\geq38}\right) = \sqrt{\dfrac{\hat{p}_{pooled}\hat{q}_{pooled}}{n_{<38}} + \dfrac{\hat{p}_{pooled}\hat{q}_{pooled}}{n_{\geq38}}} = \sqrt{\dfrac{\left(\frac{49}{246}\right)\left(\frac{197}{246}\right)}{157} + \dfrac{\left(\frac{49}{246}\right)\left(\frac{197}{246}\right)}{89}} \approx 0.0530.$

The observed difference between the proportions is:
0.2675 – 0.0787 = 0.1888.

Since the *P*-value = 0.0004 is low, we reject the null hypothesis. There is strong evidence to suggest a difference in the proportion of live births for women under 38 and women 38 and over at this clinic. In fact, the evidence suggests that women under 38 have a higher proportion of live births.

$z = \dfrac{0.1888 - 0}{0.0530}$

$z \approx 3.56$

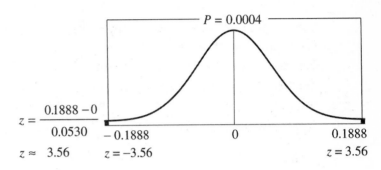

b) $\left(\hat{p}_{<38} - \hat{p}_{\geq38}\right) \pm z^{*}\sqrt{\dfrac{\hat{p}_{<38}\hat{q}_{<38}}{n_{<38}} + \dfrac{\hat{p}_{\geq38}\hat{q}_{\geq38}}{n_{\geq38}}}$

$= \left(\frac{42}{157} - \frac{7}{89}\right) \pm 1.960\sqrt{\dfrac{\left(\frac{42}{157}\right)\left(\frac{115}{157}\right)}{157} + \dfrac{\left(\frac{7}{89}\right)\left(\frac{82}{89}\right)}{89}} = (0.100, 0.278)$

We are 95% confident that the proportion of live births for patients at this clinic is between 10.0% and 27.8% higher for women under 38 than for women 38 and over. However, the Success/Failure condition is not met for the older women, so we should be cautious when using this interval. (The expected number of successes from the pooled proportion cannot be used for a condition for a confidence interval. It's based upon an assumption that the proportions are the same. We don't make that assumption in a confidence interval. In fact, we are implicitly assuming a *difference*, by finding an interval for the difference in proportion.)

15. Politics and sex.

H_0: The proportion of voters in support of the candidate is the same before and after news of his extramarital affair got out. $\left(p_B = p_A \text{ or } p_B - p_A = 0\right)$

H_A: The proportion of voters in support of the candidate has decreased after news of his extramarital affair got out. $\left(p_B > p_A \text{ or } p_B - p_A > 0\right)$

Randomization condition: Voters were randomly selected.
10% condition: 630 and 1010 are both less than 10% of all voters.
Independent samples condition: Since the samples were random, the groups are independent.
Success/Failure condition: $n\hat{p}$(before) = (630)(0.54) = 340, $n\hat{q}$(before) = (630)(0.46) = 290, $n\hat{p}$(after) = (1010)(0.51) = 515, and $n\hat{q}$(after) = (1010)(0.49) = 505 are all greater than 10, so both samples are large enough.

Since the conditions have been satisfied, we will model the sampling distribution of the difference in proportion with a Normal model with mean 0 and standard deviation estimated by:

$$SE_{pooled}(\hat{p}_B - \hat{p}_A) = \sqrt{\frac{\hat{p}_{pooled}\hat{q}_{pooled}}{n_B} + \frac{\hat{p}_{pooled}\hat{q}_{pooled}}{n_A}} = \sqrt{\frac{(0.5215)(0.4785)}{630} + \frac{(0.5215)(0.4785)}{1010}} \approx 0.02536.$$

The observed difference between the proportions is:
0.54 – 0.51 = 0.03.

Since the *P*-value = 0.118 is fairly high, we fail to reject the null hypothesis. There is little evidence of a decrease in the proportion of voters in support of the candidate after the news of his extramarital affair got out.

$$z = \frac{0.03 - 0}{0.02536}$$

$$z \approx 1.18$$

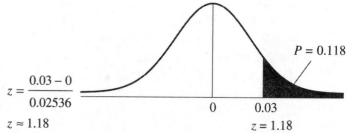

17. Twins.

a) H_0: The proportion of multiple births is the same for white women and black women.
$(p_W = p_B$ or $p_W - p_B = 0)$

H_A : The proportion of multiple births is different for white women and black women.
$(p_W \neq p_B$ or $p_W - p_B \neq 0)$

Randomization condition: Assume that these women are representative of all women.
10% condition: 3132 and 606 are both less than 10% of all people.
Independent samples condition: The groups are independent.
Success/Failure condition: $n\hat{p}$ (white) = 94, $n\hat{q}$ (white) = 3038, $n\hat{p}$ (black) = 20, and $n\hat{q}$ (black) = 586 are all greater than 10, so both samples are large enough.

Since the conditions have been satisfied, we will model the sampling distribution of the difference in proportion with a Normal model with mean 0 and standard deviation estimated by:

$$SE_{pooled}(\hat{p}_W - \hat{p}_B) = \sqrt{\frac{\hat{p}_{pooled}\hat{q}_{pooled}}{n_W} + \frac{\hat{p}_{pooled}\hat{q}_{pooled}}{n_B}} = \sqrt{\frac{\left(\frac{114}{3738}\right)\left(\frac{3624}{3738}\right)}{3132} + \frac{\left(\frac{114}{3738}\right)\left(\frac{3624}{3738}\right)}{606}} \approx 0.007631.$$

The observed difference between the proportions is:
0.030 – 0.033 = –0.003.

Since the *P*-value = 0.6951 is high, we fail to reject the null hypothesis. There is no evidence of a difference between the proportions of multiple births for white women and black women.

$$z = \frac{0.003 - 0}{0.07631}$$

$$z \approx -0.39$$

b)   If there is actually a difference between the proportions of multiple births for white women and black women, then we have committed a Type II error.

19. Mammograms.

a) H_0: The proportion of deaths from breast cancer is the same for women who never had a mammogram as for women who had mammograms. $(p_N = p_M$ or $p_N - p_M = 0)$

H_A: The proportion of deaths from breast cancer is greater for women who never had a mammogram than for women who had mammograms. $(p_N > p_M$ or $p_N - p_M > 0)$

Randomization condition: Assume that the women are representative of all women.
10% condition: 30,565 and 30,131 are both less than 10% of all women.
Independent samples condition: The groups were chosen independently.
Success/Failure condition: $n\hat{p}$ (never) = 196, $n\hat{q}$ (never) = 30,369, $n\hat{p}$ (mammogram) = 153, and $n\hat{q}$ (mammogram) = 29,978 are all greater than 10, so both samples are large enough.

Since the conditions have been satisfied, we will model the sampling distribution of the difference in proportion with a Normal model with mean 0 and standard deviation estimated by:

$$SE_{pooled}(\hat{p}_N - \hat{p}_M) = \sqrt{\frac{\hat{p}_{pooled}\hat{q}_{pooled}}{n_N} + \frac{\hat{p}_{pooled}\hat{q}_{pooled}}{n_M}} = \sqrt{\frac{\left(\frac{349}{60,696}\right)\left(\frac{60,347}{60,696}\right)}{30,565} + \frac{\left(\frac{349}{60,696}\right)\left(\frac{60,347}{60,696}\right)}{30,131}} \approx 0.0006138.$$

The observed difference between the proportions is: 0.006413 – 0.005078 = 0.001335.

Since the *P*-value = 0.0148 is low, we reject the null hypothesis. There is strong evidence that the proportion of breast cancer deaths for women who have never had a mammogram is greater than the proportion of deaths from breast cancer for women who underwent screening by mammogram.

$$z = \frac{0.001335 - 0}{0.0006198}$$

$$z \approx 2.17$$

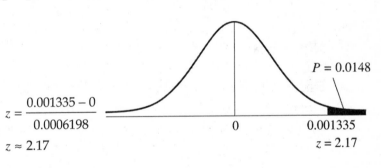

b) If there is actually no difference between the proportions of deaths from breast cancer for women who have and have not had mammograms, we have committed a Type I error.

21. Pain.

a) **Randomization condition:** The patients were randomly selected AND randomly assigned to treatment groups. If that's not random enough for you, I don't know what is!
10% condition: 112 and 108 are both less than 10% of all people with joint pain.
Success/Failure condition: $n\hat{p}(A) = 84$, $n\hat{q}(A) = 28$, $n\hat{p}(B) = 66$, and $n\hat{q}(B) = 42$ are all greater than 10, so both samples are large enough.

Since the conditions are met, we can use a one-proportion z-interval to estimate the percentage of patients who may get relief from medication A.

$$\hat{p} \pm z^* \sqrt{\frac{\hat{p}\hat{q}}{n}} = \left(\frac{84}{112}\right) \pm 1.960 \sqrt{\frac{\left(\frac{84}{112}\right)\left(\frac{28}{112}\right)}{112}} = (67.0\%, 83.0\%)$$

We are 95% confident that between 67.0% and 83.0% of patients with joint pain will find medication A to be effective.

b) Since the conditions were met in part a, we can use a one-proportion z-interval to estimate the percentage of patients who may get relief from medication B.

$$\hat{p} \pm z^* \sqrt{\frac{\hat{p}\hat{q}}{n}} = \left(\frac{66}{108}\right) \pm 1.960 \sqrt{\frac{\left(\frac{66}{108}\right)\left(\frac{42}{108}\right)}{108}} = (51.9\%, 70.3\%)$$

We are 95% confident that between 51.9% and 70.3% of patients with joint pain will find medication B to be effective.

c) The 95% confidence intervals overlap, which might lead one to believe that there is no evidence of a difference in the proportions of people who find each medication effective. However, if one was lead to believe that, one should proceed to part…

d) Most of the conditions were checked in part a. We only have one more to check:
Independent samples condition: The groups were assigned randomly, so there is no reason to believe there is a relationship between them.

Since the conditions have been satisfied, we will find a two-proportion z-interval.

$$\left(\hat{p}_A - \hat{p}_B\right) \pm z^* \sqrt{\frac{\hat{p}_A\hat{q}_A}{n_A} + \frac{\hat{p}_B\hat{q}_B}{n_B}}$$

$$= \left(\tfrac{84}{112} - \tfrac{66}{108}\right) \pm 1.960 \sqrt{\frac{\left(\frac{84}{112}\right)\left(\frac{28}{112}\right)}{112} + \frac{\left(\frac{66}{108}\right)\left(\frac{42}{108}\right)}{112}} = (0.017, 0.261)$$

We are 95% confident that the proportion of patients with joint pain who will find medication A effective is between 1.70% and 26.1% higher than the proportion of patients who will find medication B effective.

e) The interval does not contain zero. There is evidence that medication A is more effective than medication B.

f) The two-proportion method is the proper method. By attempting to use two, separate, confidence intervals, you are adding standard deviations when looking for a difference in proportions. We know from our previous studies that *variances* add when finding the standard deviation of a difference. The two-proportion method does this.

Review of Part V – From the Data at Hand to the World at Large

1. **Herbal cancer.**

   H_0: The cancer rate for those taking the herb is the same as the cancer rate for those not taking the herb. $\left(p_{Herb} = p_{Not} \text{ or } p_{Herb} - p_{Not} = 0 \right)$

   H_A: The cancer rate for those taking the herb is higher than the cancer rate for those not taking the herb. $\left(p_{Herb} > p_{Not} \text{ or } p_{Herb} - p_{Not} > 0 \right)$

3. **Birth days.**

   a) If births are distributed uniformly across all days, we expect the number of births on each day to be $np = (72)(\frac{1}{7}) \approx 10.29$.

   b) **10% condition:** The 72 births are likely to be representative of all births at the hospital with regards to day of birth, and 72 births are less than 10% of the births.
   Success/Failure condition: The expected number of births on a particular day of the week is $np = (72)(\frac{1}{7}) \approx 10.29$ and the expected number of births not on that particular day is $nq = (72)(\frac{6}{7}) \approx 61.71$. These are both greater than 10, so the sample is large enough.

   Since the conditions have been satisfied, a Normal model can be used to model the sampling distribution of the proportion of 72 births that occur on a given day of the week.

   $\mu_{\hat{p}} = p = \frac{1}{7} \approx 0.1429$

   $\sigma(\hat{p}) = \sqrt{\dfrac{pq}{n}} = \sqrt{\dfrac{(\frac{1}{7})(\frac{6}{7})}{72}} \approx 0.04124$

   There were 7 births on Mondays, so $\hat{p} = \dfrac{7}{72} \approx 0.09722$. This is only about a 1.11 standard deviations below the expected proportion, so there's no evidence that this is an unusual occurrence.

   c) The 17 births on Tuesdays represent an unusual occurrence. For Tuesdays, $\hat{p} = \dfrac{17}{72} \approx 0.2361$, which is about 2.26 standard deviations above the expected proportion of births. There is evidence to suggest that the proportion of births on Tuesdays is higher than expected, if births are distributed uniformly across days.

   d) Some births are scheduled for the convenience of the doctor and/or the mother.

5. **Leaky gas tanks.**

   a) H_0: The proportion of leaky gas tanks is 40%. $(p = 0.40)$
   H_A: The proportion of leaky gas tanks is less than 40%. $(p < 0.40)$

   b) **Randomization condition:** A random sample of 27 service stations in California was taken.
   10% condition: 27 service stations are less than 10% of all service stations in California.
   Success/Failure condition: $np = (27)(0.40) = 10.8$ and $nq = (27)(0.60) = 16.2$ are both greater than 10, so the sample is large enough.

c) Since the conditions have been satisfied, a Normal model can be used to model the sampling distribution of the proportion, with $\mu_{\hat{p}} = p = 0.40$ and

$$\sigma(\hat{p}) = \sqrt{\frac{pq}{n}} = \sqrt{\frac{(0.40)(0.60)}{27}} \approx 0.09428.$$ We can perform a one-proportion z-test.

The observed proportion of leaky gas tanks is $\hat{p} = \dfrac{7}{27} \approx 0.2593$.

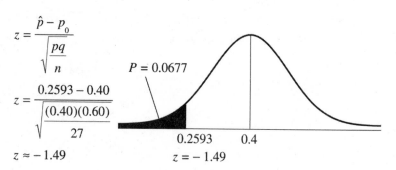

$$z = \frac{\hat{p} - p_0}{\sqrt{\dfrac{pq}{n}}}$$

$$z = \frac{0.2593 - 0.40}{\sqrt{\dfrac{(0.40)(0.60)}{27}}}$$

$$z \approx -1.49$$

d) Since the P-value = 0.0677 is relatively high, we fail to reject the null hypothesis. There is little evidence that the proportion of leaky gas tanks is less than 40%. The new program doesn't appear to be effective in decreasing the proportion of leaky gas tanks.

e) If the program actually works, we haven't done anything *wrong*. Our methods are correct. Statistically speaking, we have committed a Type II error.

f) In order to decrease the probability of making this type of error, we could lower our standards of proof, by raising the level of significance. This will increase the power of the test to detect a decrease in the proportion of leaky gas tanks. Another way to decrease the probability that we make a Type II error is to sample more service stations. This will decrease the variation in the sample proportion, making our results more reliable.

g) Increasing the level of significance is advantageous, since it decreases the probability of making a Type II error, and increases the power of the test. However, it also increases the probability that a Type I error is made, in this case, thinking that the program is effective when it really is not effective.
Increasing the sample size decreases the probability of making a Type II error and increases power, but can be costly and time-consuming.

7. Scrabble.

a) The researcher believes that the true proportion of As is within 10% of the estimated 54%, namely, between 44% and 64%.

b) A large margin of error is usually associated with a small sample, but the sample consisted of "many" hands. The margin of error is large because the standard error of the sample is large. This occurs because the true proportion of As in a hand is close to 50%, the most difficult proportion to predict.

c) This provides no evidence that the simulation is faulty. The true proportion of As is contained in the confidence interval. The researcher's results are consistent with 63% As.

9. News sources.

a) The Pew Research Foundation believes that the true proportion of people who obtain news from the Internet is between 30% and 36%.

b) The smaller sample size in the limited sample would result in a larger standard error. This would make the margin of error larger, as well.

c) $\hat{p}\pm z^*\sqrt{\dfrac{\hat{p}\hat{q}}{n}}=(0.45)\pm1.960\sqrt{\dfrac{(0.45)(0.55)}{239}}=(38.7\%,51.3\%)$

We are 95% confident that between 38.7% and 51.3% of active traders rely on the Internet for investment information.

d) The sample of 239 active traders is smaller than either of the earlier samples. This results in a larger margin of error.

11. Bimodal.

a) The *sample's* distribution (NOT the *sampling* distribution), is expected to look more and more like the distribution of the population, in this case, bimodal.

b) The expected value of the sample's mean is expected to be μ, the population mean, regardless of sample size.

c) The variability of the sample mean, $\sigma(\bar{y})$, is $\dfrac{\sigma}{\sqrt{n}}$, the population standard deviation divided by the square root of the sample size, regardless of the sample size.

d) As the sample size increases, the sampling distribution model becomes closer and closer to a Normal model.

13. Archery.

a) $\mu_{\hat{p}}=p=0.80$

$\sigma(\hat{p})=\sqrt{\dfrac{pq}{n}}=\sqrt{\dfrac{(0.80)(0.20)}{200}}\approx0.028$

b) $np=(200)(0.80)=160$ and $nq=(200)(0.20)=40$ are both greater than 10, so the Normal model is appropriate.

c) The Normal model of the sampling distribution of the proportion of bull's-eyes she makes out of 200 is at the right.

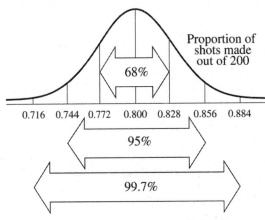

Approximately 68% of the time, we expect her to hit the bull's-eye on between 77.2% and 82.8% of her shots. Approximately 95% of the time, we expect her to hit the bull's-eye on between 74.4% and 85.6% of her shots. Approximately 99.7% of the time, we expect her to hit the bull's-eye on between 71.6% and 88.4% of her shots.

d) According to the Normal model, the probability that she hits the bull's-eye in at least 85% of her 200 shots is approximately 0.039.

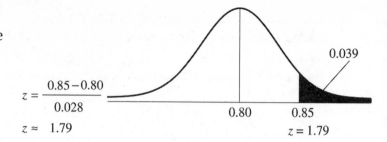

$$z = \frac{0.85 - 0.80}{0.028}$$

$$z \approx 1.79$$

15. Twins.

H_0: The proportion of preterm twin births in 1990 is the same as the proportion of preterm twin births in 2000. $\left(p_{1990} = p_{2000} \text{ or } p_{1990} - p_{2000} = 0\right)$

H_A: The proportion of preterm twin births in 1990 is the less than the proportion of preterm twin births in 2000. $\left(p_{1990} < p_{2000} \text{ or } p_{1990} - p_{2000} < 0\right)$

Randomization condition: Assume that these births are representative of all twin births.
10% condition: 43 and 48 are both less than 10% of all twin births.
Independent samples condition: The samples are from different years, so they are unlikely to be related.
Success/Failure condition: $n\hat{p}\,(1990) = 20$, $n\hat{q}\,(1990) = 23$, $n\hat{p}\,(2000) = 26$, and $n\hat{q}\,(2000) = 22$ are all greater than 10, so both samples are large enough.

Since the conditions have been satisfied, we will perform a two-proportion z-test. We will model the sampling distribution of the difference in proportion with a Normal model with mean 0 and standard deviation estimated by:

$$SE_{pooled}\left(\hat{p}_{1990} - \hat{p}_{2000}\right) = \sqrt{\frac{\hat{p}_{pooled}\hat{q}_{pooled}}{n_{1900}} + \frac{\hat{p}_{pooled}\hat{q}_{pooled}}{n_{2000}}} = \sqrt{\frac{\left(\frac{46}{91}\right)\left(\frac{45}{91}\right)}{43} + \frac{\left(\frac{46}{91}\right)\left(\frac{45}{91}\right)}{48}} \approx 0.1050.$$

The observed difference between the proportions is:
0.4651 – 0.5417 = – 0.0766

Since the *P*-value = 0.2329 is high, we fail to reject the null hypothesis. There is no evidence of an increase in the proportion of preterm twin births from 1990 to 2000, at least not at this large city hospital.

$$z = \frac{-0.0766 - 0}{0.1050}$$

$$z \approx -0.73$$

17. Eclampsia.

a) H_0: The proportion of pregnant women who die after developing eclampsia is the same for women taking magnesium sulfide as it is for women not taking magnesium sulfide. $\left(p_{MS} = p_N \text{ or } p_{MS} - p_N = 0\right)$

H_A: The proportion of pregnant women who die after developing eclampsia is lower for women taking magnesium sulfide than for women not taking magnesium sulfide. $\left(p_{MS} < p_N \text{ or } p_{MS} - p_N < 0\right)$

b) **Randomization condition:** Although not specifically stated, these results are from a large-scale experiment, which was undoubtedly properly randomized.
10% condition: 40 and 96 are less than 10% of all pregnant women.
Independent samples condition: Subjects were randomly assigned to the treatments.
Success/Failure condition: $n\hat{p}$ (mag. sulf.) = 11, $n\hat{q}$ (mag. sulf.) = 29, $n\hat{p}$ (placebo) = 20, and $n\hat{q}$ (placebo) = 76 are all greater than 10, so both samples are large enough.

Since the conditions have been satisfied, we will perform a two-proportion z-test. We will model the sampling distribution of the difference in proportion with a Normal model with mean 0 and standard deviation estimated by:

$$SE_{pooled}(\hat{p}_{MS} - \hat{p}_N) = \sqrt{\frac{\hat{p}_{pooled}\hat{q}_{pooled}}{n_{MS}} + \frac{\hat{p}_{pooled}\hat{q}_{pooled}}{n_N}} = \sqrt{\frac{\left(\frac{31}{136}\right)\left(\frac{105}{136}\right)}{40} + \frac{\left(\frac{31}{136}\right)\left(\frac{105}{136}\right)}{96}} \approx 0.07895.$$

c) The observed difference between the proportions is:
$0.275 - 0.2083 = 0.0667$

Since the P-value = 0.8008 is high, we fail to reject the null hypothesis. There is no evidence that the proportion of women who may die after developing eclampsia is lower for women taking magnesium sulfide than for women who are not taking the drug.

$z = \dfrac{0.0667 - 0}{0.07895}$

$z \approx 0.84$

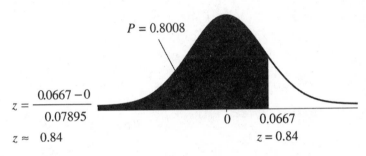

$P = 0.8008$

0   0.0667
$z = 0.84$

d) There is not sufficient evidence to conclude that magnesium sulfide is effective in preventing death when eclampsia develops.

e) If magnesium sulfide is effective in preventing death when eclampsia develops, then we have made a Type II error.

f) To increase the power of the test to detect a decrease in death rate due to magnesium sulfide, we could increase the sample size or increase the level of significance.

g) Increasing the sample size lowers variation in the sampling distribution, but may be costly. The sample size is already quite large. Increasing the level of significance increases power by increasing the likelihood of rejecting the null hypothesis, but increases the chance of making a Type I error, namely thinking that magnesium sulfide is effective when it is not.

19. Polling disclaimer.

a) It is not clear what specific question the pollster asked. Otherwise, they did a great job of identifying the W's.

b) A sample that was stratified by age, sex, region, and education was used.

c) The margin of error was 4%.

d) Since "no more than 1 time in 20 should chance variations in the sample cause the results to vary by more than 4 percentage points", the confidence level is 19/20 = 95%.

e) The subgroups had smaller sample sizes than the larger group. The standard errors in these subgroups were larger as a result, and this caused the margins of error to be larger.

f) They cautioned readers about response bias due to wording and order of the questions.

21. Teen deaths.

a) H_0 : The percentage of fatal accidents involving teenage girls is 14.3%, the same as the overall percentage of fatal accidents involving teens . ($p = 0.143$)
H_A : The percentage of fatal accidents involving teenage girls is lower than 14.3%, the overall percentage of fatal accidents involving teens . ($p < 0.143$)

Plausible independence: It is reasonable to think that accidents occur independently.
Randomization condition: Assume that the 388 accidents observed are representative of all accidents.
10% condition: The sample of 388 accidents is less than 10% of all accidents.
Success/Failure condition: $np = (388)(0.143) = 55.484$ and $nq = (388)(0.857) = 332.516$ are both greater than 10, so the sample is large enough.

The conditions have been satisfied, so a Normal model can be used to model the sampling distribution of the proportion, with $\mu_{\hat{p}} = p = 0.143$ and

$$\sigma(\hat{p}) = \sqrt{\frac{pq}{n}} = \sqrt{\frac{(0.143)(0.857)}{388}} \approx 0.01777 .$$

We can perform a one-proportion z-test. The observed proportion of fatal accidents involving teen girls is $\hat{p} = \dfrac{44}{388} \approx 0.1134$.

Since the *P*-value = 0.0479 is low, we reject the null hypothesis. There is some evidence that the proportion of fatal accidents involving teen girls is less than the overall proportion of fatal accidents involving teens.

$$z = \frac{\hat{p} - p_0}{\sqrt{\dfrac{pq}{n}}}$$

$$z = \frac{0.1134 - 0.143}{\sqrt{\dfrac{(0.143)(0.857)}{388}}}$$

$$z \approx -1.67$$

$P = 0.0479$

0.1134   0.143
$z = -1.67$

b) If the proportion of fatal accidents involving teenage girls is really 14.3%, we expect to see the observed proportion, 11.34%, in about 4.79% of samples of size 388 simply due to sampling variation.

23. Largemouth bass.

a) One would expect many small fish, with a few large fish.

b) We cannot determine the probability that a largemouth bass caught from the lake weighs over 3 pounds because we don't know the exact shape of the distribution. We know that it is NOT Normal.

c) It would be quite risky to attempt to determine whether or not the mean weight of 5 fish was over 3 pounds. With a skewed distribution, a sample of size 5 is not large enough for the Central Limit Theorem to guarantee that a Normal model is appropriate to describe the distribution of the mean.

d) A sample of 60 randomly selected fish is large enough for the Central Limit Theorem to guarantee that a Normal model is appropriate to describe the sampling distribution of the mean, as long as 60 fish is less than 10% of the population of all the fish in the lake.

The mean weight is $\mu = 3.5$ pounds, with standard deviation $\sigma = 2.2$ pounds. Since the sample size is sufficiently large, we can model the sampling distribution of the mean weight of 60 fish with a Normal model, with $\mu_{\bar{y}} = 3.5$ pounds and standard deviation

$$\sigma(\bar{y}) = \frac{2.2}{\sqrt{60}} \approx 0.284 \text{ pounds}.$$

According to the Normal model, the probability that 60 randomly selected fish average more than 3 pounds is approximately 0.961.

$$z = \frac{3 - 3.5}{\frac{2.2}{\sqrt{60}}}$$

$$z \approx -1.76$$

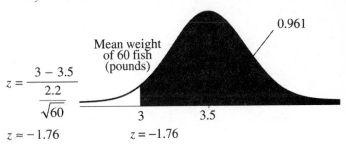

25. Language.

a) **10% condition:** The 60 people selected randomly represent less than 10% of all people. **Success/Failure condition:** $np = (60)(0.80) = 48$ and $nq = (60)(0.20) = 12$ are both greater than 10.

Therefore, the sampling distribution model for the proportion of 60 randomly selected people who have left-brain language control is Normal, with $\mu_{\hat{p}} = p = 0.80$ and standard

deviation $\sigma(\hat{p}) = \sqrt{\dfrac{pq}{n}} = \sqrt{\dfrac{(0.80)(0.20)}{60}} \approx 0.0516$.

b) According to the Normal model, the probability that over 75% of these 60 people have left-brain language control is approximately 0.894.

$$z = \frac{\hat{p} - \mu_{\hat{p}}}{\sqrt{\dfrac{pq}{n}}}$$

$$z = \frac{0.75 - 0.80}{\sqrt{\dfrac{(0.80)(0.20)}{60}}}$$

$$z \approx -0.968$$

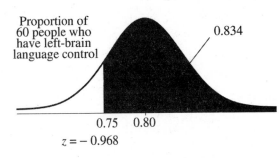

c) If the sample had consisted of 100 people, the probability would have been higher. A larger sample results in a smaller standard deviation for the sample proportion.

d) Answers may vary. Let's consider three standard deviations below the expected proportion to be "almost certain". It would take a sample of (exactly!) 576 people to make sure that 75% would be 3 standard deviations below the expected percentage of people with left-brain language control.

Using round numbers for n instead of z, about 500 people in the sample would make the probability of choosing a sample with at least 75% of the people having left-brain language control is a whopping 0.997.

It all depends on what "almost certain" means to you.

$$z = \frac{\hat{p} - \mu_{\hat{p}}}{\sqrt{\dfrac{pq}{n}}}$$

$$-3 = \frac{0.75 - 0.80}{\sqrt{\dfrac{(0.80)(0.20)}{n}}}$$

$$n = \frac{(-3)^2 (0.80)(0.20)}{(0.75 - 0.80)^2} = 576$$

27. Crohn's disease.

a) **Plausible independence:** It is reasonable to think that the patients would respond to infliximab independently of each other.
Randomization condition: Assume that the 573 patients are representative of all Crohn's disease sufferers.
10% condition: 573 patients are less than 10% of all sufferers of Crohn's disease.
Success/Failure condition: $n\hat{p} = 335$ and $n\hat{q} = 238$ are both greater than 10, so the sample is large enough.

Since the conditions are met, we can use a one-proportion z-interval to estimate the percentage of Crohn's disease sufferers who respond positively to infliximab.

$$\hat{p} \pm z^* \sqrt{\frac{\hat{p}\hat{q}}{n}} = \left(\frac{335}{573}\right) \pm 1.960 \sqrt{\frac{\left(\frac{335}{573}\right)\left(\frac{238}{573}\right)}{573}} = (54.5\%, 62.5\%)$$

b) We are 95% confident that between 54.5% and 62.5% of Crohn's disease sufferers would respond positively to infliximab.

c) 95% of random samples of size 573 will produce intervals that contain the true proportion of Crohn's disease sufferers who respond positively to infliximab.

29. Alcohol abuse.

$$ME = z^* \sqrt{\frac{\hat{p}\hat{q}}{n}}$$

$$0.04 = 1.645 \sqrt{\frac{(0.5)(0.5)}{n}}$$

$$n = \frac{(1.645)^2(0.5)(0.5)}{(0.04)^2}$$

$$n \approx 423$$

The university will have to sample at least 423 students in order to estimate the proportion of students who have been drunk with in the past week to within $\pm 4\%$, with 90% confidence.

31. Preemies.

a) **Randomization condition:** Assume that these kids are representative of all kids.
10% condition: 242 and 233 are less than 10% of all kids.
Independent samples condition: The groups are independent.
Success/Failure condition: $n\hat{p}$ (preemies) = (242)(0.74) = 179, $n\hat{q}$ (preemies) = (242)(0.26) = 63, $n\hat{p}$ (normal weight) = (233)(0.83) = 193, and $n\hat{q}$ (normal weight) = 40 are all greater than 10, so the samples are both large enough.

Since the conditions have been satisfied, we will find a two-proportion z-interval.

$$(\hat{p}_N - \hat{p}_P) \pm z^* \sqrt{\frac{\hat{p}_N\hat{q}_N}{n_N} + \frac{\hat{p}_P\hat{q}_P}{n_P}} = (0.83 - 0.74) \pm 1.960 \sqrt{\frac{(0.83)(0.17)}{233} + \frac{(0.74)(0.26)}{242}} = (0.017, 0.163)$$

We are 95% confident that between 1.7% and 16.3% more normal birth-weight children graduated from high school than children who were born premature.

b) Since the interval for the difference in percentage of high school graduates is above 0, there is evidence normal birth-weight children graduate from high school at a greater rate than premature children.

c) If preemies do not have a lower high school graduation rate than normal birth-weight children, then we made a Type I error. We rejected the null hypothesis of "no difference" when we shouldn't have.

33. Fried PCs.

a) H_0: The computer is undamaged.
H_A: The computer is damaged.

b) The biggest advantage is that all of the damaged computers will be detected, since, historically, damaged computers never pass all the tests. The disadvantage is that only 80% of undamaged computers pass all the tests. The engineers will be classifying 20% of the undamaged computers as damaged.

c) In this example, a Type I error is rejecting an undamaged computer. To allow this to happen only 5% of the time, the engineers would reject any computer that failed 3 or more tests, since 95% of the undamaged computers fail two or fewer tests.

d) The power of the test in part c is 20%, since only 20% of the damaged machines fail 3 or more tests.

e) By declaring computers "damaged" if the fail 2 or more tests, the engineers will be rejecting only 7% of undamaged computers. From 5% to 7% is an increase of 2% in α. Since 90% of the damaged computers fail 2 or more tests, the power of the test is now 90%, a substantial increase.

Chapter 23 – Inferences About Means

1. ***t*-models, part I.**

    a) 1.74   **b)** 2.37   **c)** 0.0524   **d)** 0.0889

3. ***t*-models, part III.**

    As the number of degrees of freedom increases, the shape and center of *t*-models do not change. The spread of *t*-models decreases as the number of degrees of freedom increases.

5. **Cattle.**

    a) Not correct. A confidence interval is for the mean weight gain of the population of all cows. It says nothing about individual cows. This interpretation also appears to imply that there is something special about the interval that was generated, when this interval is actually one of many that could have been generated, depending on the cows that were chosen for the sample.

    b) Not correct. A confidence interval is for the mean weight gain of the population of all cows, not individual cows.

    c) Not correct. We don't need a confidence interval about the average weight gain for cows in this study. We are certain that the mean weight gain of the cows in this study is 56 pounds. Confidence intervals are for the mean weight gain of the population of all cows.

    d) Not correct. This statement implies that the average weight gain varies. It doesn't. We just don't know what it is, and we are trying to find it. The average weight gain is either between 45 and 67 pounds, or it isn't.

    e) Not correct. This statement implies that there is something special about our interval, when this interval is actually one of many that could have been generated, depending on the cows that were chosen for the sample. The correct interpretation is that 95% of samples of this size will produce an interval that will contain the mean weight gain of the population of all cows.

7. **Pulse rates.**

    a) We are 95% confident that the mean pulse rate of adults is between 70.9 and 74.5 beats per minute.

    b) The width of the interval is about 74.5 – 70.9 = 3.6 beats per minute. The margin of error is half of that, about 1.8 beats per minute.

    c) The margin of error would have been larger. More confidence requires a larger critical value of *t*, which increases the margin of error.

9. **Normal temperature.**

a) **Randomization condition:** The adults were randomly selected.
10% condition: 52 adults are less than 10% of all adults.
Nearly Normal condition: The sample of 52 adults is large, and the histogram shows no serious skewness, outliers, or multiple modes.

The people in the sample had a mean temperature of 98.2846° and a standard deviation in temperature of 0.682379°. Since the conditions are satisfied, the sampling distribution of the mean can be modeled by a Student's t model, with 52 – 1 = 51 degrees of freedom. We will use a one-sample t-interval with 98% confidence for the mean body temperature. (By hand, use $t_{50}^* \approx 2.403$ from the table.)

b) $\bar{y} \pm t_{n-1}^* \left(\dfrac{s}{\sqrt{n}} \right) = 98.2846 \pm t_{51}^* \left(\dfrac{0.682379}{\sqrt{52}} \right) \approx (98.06, 98.51)$

c) We are 98% confident that the mean body temperature for adults is between 98.06°F and 98.51°F. (If you calculated the interval by hand, using $t_{50}^* \approx 2.403$ from the table, your interval may be slightly different than intervals calculated using technology. With the rounding used here, they are identical. Even if they aren't, it's not a big deal.)

d) 98% of all random samples of size 52 will produce intervals that contain the true mean body temperature of adults.

e) Since the interval is completely below the body temperature of 98.6°F, there is strong evidence that the true mean body temperature of adults is lower than 98.6°F.

11. **Normal temperatures, part II.**

a) The 90% confidence interval would be narrower than the 98% confidence interval. We can be more precise with our interval when we are less confident.

b) The 98% confidence interval has a greater chance of containing the true mean body temperature of adults than the 90% confidence interval, but the 98% confidence interval is less precise (wider) than the 90% confidence interval.

c) The 98% confidence interval would be narrower if the sample size were increased from 52 people to 500 people. The smaller standard error would result in a smaller margin of error.

d) Our sample of 52 people gave us a 98% confidence interval with a margin of error of (98.51 – 98.05)/2 = 0.225°F. In order to get a margin of error of 0.1, less than half of that, we need a sample over 4 times as large. It should be safe to use $t_{100}^* \approx 2.364$ from the table, since the sample will need to be larger than 101. Or we could use $z^* \approx 2.326$, since we expect the sample to be large. The important thing is to pick *something*, and go with it. This is just an *estimate*! We need a sample of about 252 people in order to estimate the mean body temperature of adults to within 0.1°F.

$$ME = t_{n-1}^* \left(\frac{s}{\sqrt{n}} \right)$$

$$0.1 = 2.326 \left(\frac{0.682379}{\sqrt{n}} \right)$$

$$n = \frac{(2.326)^2 (0.682379)^2}{(0.1)^2}$$

$$n \approx 252$$

13. Hot dogs.

a) $\bar{y} \pm t^*_{n-1}\left(\dfrac{s}{\sqrt{n}}\right) = 310 \pm t^*_{39}\left(\dfrac{36}{\sqrt{40}}\right) \approx (298.5, 321.5)$

b) We have assumed that the hot dog weights are independent and that the distribution of the population of hot dog weights is Normal. The conditions to check are:
Randomization condition: We don't know that the hot dogs were sampled at random, but it is reasonable to think that the hot dogs are representative of hot dogs of this type.
10% condition: 40 hot dogs are less than 10% of all hot dogs.
Nearly Normal condition: We don't have the actual data, but since the sample of 40 hot dogs is fairly large, it is okay to proceed.

c) We are 95% confident that the mean sodium content in this type of "reduced sodium" hot dogs is between 298.5 and 321.5 mg.

15. Second dog.

a) This larger sample should produce a more accurate estimate of the mean sodium content in the hot dogs. A larger sample has a smaller standard error, which results in a smaller margin of error.

b) $SE(\bar{y}) = \left(\dfrac{s}{\sqrt{n}}\right) = \left(\dfrac{32}{\sqrt{60}}\right) \approx 4.1$ mg sodium.

c) **Randomization condition:** We don't know that the hot dogs were sampled at random, but it is reasonable to think that the hot dogs are representative of hot dogs of this type.
10% condition: 60 hot dogs are less than 10% of all hot dogs.
Nearly Normal condition: We don't have the actual data, but since the sample of 60 hot dogs is large it is okay to proceed.

The hot dogs in the sample had a mean sodium content of 318 mg, and a standard deviation in sodium content of 32 mg. Since the conditions have been satisfied, construct a one-sample t-interval, with $60 - 1 = 59$ degrees of freedom, at 95% confidence.

$\bar{y} \pm t^*_{n-1}\left(\dfrac{s}{\sqrt{n}}\right) = 318 \pm t^*_{59}\left(\dfrac{32}{\sqrt{60}}\right) \approx (309.7, 326.3)$

We are 95% confident that the mean sodium content of this type of hot dog is between 309.7 and 326.3 mg.

d) If a "reduced sodium" hot dog has to have at least 30% less sodium than a hot dog containing an average of 465 mg of sodium, then it must have less than 0.70(465) = 325.5 mg of sodium. Since this value is contained in our 95% confidence interval, there is little evidence to suggest that this type of hot dog can be labeled as "reduced sodium".

17. TV Safety.

a) The inspectors are performing an upper-tail test. They need to prove that the stands will support 500 pounds (or more) easily.

b) The inspectors commit a Type I error if they certify the stands as safe, when they are not.

c) The inspectors commit a Type II error if they decide the stands are not safe, when they are.

19. TV safety revisited.

a) The value of α should be decreased. This means a smaller chance of declaring the stands safe under the null hypothesis that they are not safe.

b) The power of the test is the probability of correctly detecting that the stands can safely hold over 500 pounds.

c) 1) The company could redesign the stands so that their strength is more consistent, as measured by the standard deviation. Redesigning the manufacturing process is likely to be quite costly.
2) The company could increase the number of stands tested. This costs them both time to perform the test and money to pay the quality control personnel.
3) The company could increase α, effectively lowering their standards for what is required to certify the stands "safe". This is a big risk, since there is a greater chance of Type I error, namely allowing unsafe stands to be sold.
4) The company could make the stands stronger, increasing the mean amount of weight that the stands can safely hold. This type of redesign is expensive.

21. Marriage.

a) H_0: The mean age at which American men first marry is 23.3 years. $(\mu = 23.3)$
H_A: The mean age at which American men first marry is greater than 23.3 years. $(\mu > 23.3)$

b) **Randomization condition:** The 40 men were selected randomly.
10% condition: 40 men are less than 10% of all recently married men.
Nearly Normal condition: The population of ages of men at first marriage is likely to be skewed to the right. It is much more likely that there are men who marry for the first time at an older age than at an age that is very young. We should examine the distribution of the sample to check for serious skewness and outliers, but with a large sample of 40 men, it should be safe to proceed.

c) Since the conditions for inference are satisfied, we can model the sampling distribution of the mean age of men at first marriage with $N\left(23.3, \dfrac{\sigma}{\sqrt{n}}\right)$. Since we do not know σ, the standard deviation of the population, $\sigma(\bar{y})$ will be estimated by $SE(\bar{y}) = \dfrac{s}{\sqrt{n}}$, and we will use a Student's t model, with $40 - 1 = 39$ degrees of freedom, $t_{39}\left(23.3, \dfrac{s}{\sqrt{40}}\right)$.

d) The mean age at first marriage in the sample was 24.2 years, with a standard deviation in age of 5.3 years. Use a one-sample t-test, modeling the sampling distribution of $\bar{y}$ with

$$t = \frac{\bar{y} - \mu_0}{SE(\bar{y})}$$

$$t = \frac{24.2 - 23.3}{\dfrac{5.3}{\sqrt{40}}}$$

$$t \approx 1.07$$

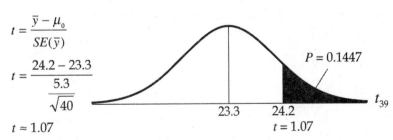

$$t_{39}\left(23.3, \frac{5.3}{\sqrt{40}}\right).$$

The *P*-value is 0.1447.

e) If the mean age at first marriage is still 23.3 years, there is a 14.5% chance of getting a sample mean of 24.2 years or older simply from natural sampling variation.

f) Since the *P*-value = 0.1447 is high, we fail to reject the null hypothesis. There is no evidence to suggest that the mean age of men at first marriage has changed from 23.3 years, the mean in 1960.

23. Ruffles.

a) **Randomization condition:** The 6 bags were not selected at random, but it is reasonable to think that these bags are representative of all bags of chips.
10% condition: 6 bags are less than 10% of all bags of chips.
Nearly Normal condition: The histogram of the weights of chips in the sample is nearly normal.

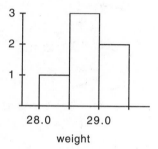

b) $\bar{y} \approx 28.78$ grams, $s \approx 0.40$ grams

c) Since the conditions for inference have been satisfied, use a one-sample *t*-interval, with 6 – 1 = 5 degrees of freedom, at 95% confidence.

$$\bar{y} \pm t^*_{n-1}\left(\frac{s}{\sqrt{n}}\right) = 28.78 \pm t^*_5\left(\frac{0.40}{\sqrt{6}}\right) \approx (28.36, 29.21)$$

d) We are 95% confident that the mean weight of the contents of Ruffles bags is between 28.36 and 29.21 grams.

e) Since the interval is above the stated weight of 28.3 grams, there is evidence that the company is filling the bags to more than the stated weight, on average.

25. Cars.

H_0: The mean weight of cars currently licensed in the U.S. is 3000 pounds. ($\mu = 3000$)
H_A: The mean weight of cars currently licensed in the U.S. is not 3000 pounds. ($\mu \neq 3000$)

Randomization condition: The 91 cars in the sample were randomly selected.
10% condition: 91 cars are less than 10% of all cars.
Nearly Normal condition: We don't have the actual data, so we cannot look at a graphical display, but since the sample is large, it is safe to proceed.

The cars in the sample had a mean weight of 2919 pounds and a standard deviation in weight of 531.5 pounds. Since the conditions for inference are satisfied, we can model the sampling distribution of the mean weight of cars currently licensed in the U.S. with a Student's *t* model, with 91 – 1 = 90 degrees of freedom, $t_{90}\left(3000, \frac{531.5}{\sqrt{91}}\right)$.

We will perform a one-sample *t*-test.

Since the *P*-value = 0.1495 is high, we fail to reject the null hypothesis. There is little evidence to suggest that the mean weight of cars in the U.S. is different than 3000 pounds.

$$t = \frac{\bar{y} - \mu_0}{SE(\bar{y})}$$

$$t = \frac{2919 - 3000}{\frac{531.5}{\sqrt{91}}}$$

$$t \approx -1.45$$

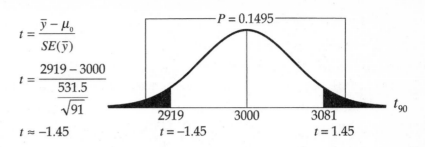

27. Chips ahoy.

a) **Randomization condition:** The bags of cookies were randomly selected.
10% condition: 16 bags are less than 10% of all bags
Nearly Normal condition: The Normal probability plot is reasonably straight, and the histogram of the number of chips per bag is unimodal and symmetric.

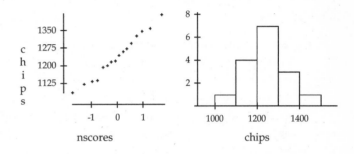

b) The bags in the sample had with a mean number of chips or 1238.19, and a standard deviation of 94.282 chips. Since the conditions for inference have been satisfied, use a one-sample *t*-interval, with 16 – 1 = 15 degrees of freedom, at 95% confidence.

$$\bar{y} \pm t_{n-1}^* \left(\frac{s}{\sqrt{n}} \right) = 1238.19 \pm t_{15}^* \left(\frac{94.282}{\sqrt{16}} \right) \approx (1187.9, \ 1288.4)$$

We are 95% confident that the mean number of chips in an 18-ounce bag of Chips Ahoy cookies is between 1187.9 and 1288.4.

c) H_0: The mean number of chips per bag is 1000. $(\mu = 1000)$

H_A: The mean number of chips per bag is greater than 1000. $(\mu > 1000)$

Since the confidence interval is well above 1000, there is strong evidence that the mean number of chips per bag is well above 1000.

However, since the "1000 Chip Challenge" is about individual bags, not means, the claim made by Nabisco may not be true. If the mean was around 1188 chips, the low end of our confidence interval, and the standard deviation of the population was about 94 chips, our best estimate obtained from our sample, a bag containing 1000 chips would be about 2 standard deviations below the mean. This is not likely to happen, but not an outrageous occurrence. These data do not provide evidence that the "1000 Chip Challenge" is true.

29. Maze.

a) **Independence assumption:** It is reasonable to think that the rats' times will be independent, as long as the times are for different rats.

Nearly Normal condition: There is an outlier in both the Normal probability plot and the histogram that should probably be eliminated before continuing the test. One rat took a long time to complete the maze.

b) H_0: The mean time for rats to complete this maze is 60 seconds. $(\mu = 60)$

H_A: The mean time for rats to complete this maze is not 60 seconds. $(\mu \neq 60)$

The rats in the sample finished the maze with a mean time of 52.21 seconds and a standard deviation in times of 13.5646 seconds. Since the conditions for inference are satisfied, we can model the sampling distribution of the mean time in which rats complete the maze with a Student's t model, with

$21 - 1 = 20$ degrees of freedom, $t_{20}\left(60, \dfrac{13.5646}{\sqrt{21}}\right)$. We will perform a one-sample t-test.

Since the P-value = 0.0160 is low, we reject the null hypothesis. There is evidence that the mean time required for rats to finish the maze is not 60 seconds. Our evidence suggests that the mean time is actually less than 60 seconds.

$$t = \frac{\bar{y} - \mu_0}{SE(\bar{y})}$$

$$t = \frac{52.21 - 60}{\dfrac{13.56}{\sqrt{21}}}$$

$$t \approx -2.63$$

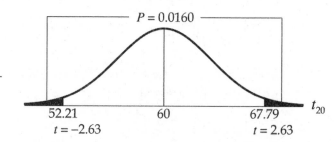

c) Without the outlier, the rats in the sample finished the maze with a mean time of 50.13 seconds and standard deviation in times of 9.90 seconds. Since the conditions for inference are satisfied, we can model the sampling distribution of the mean time in which rats complete the maze with a Student's t model, with $20 - 1 = 19$ degrees of freedom, $t_{19}\left(60, \dfrac{9.90407}{\sqrt{20}}\right)$. We will perform a one-sample t-test.

This test results in a value of $t = -4.46$, and a two-sided P-value = 0.0003. Since the P-value is low, we reject the null hypothesis. There is evidence that the mean time required for rats to finish the maze is not 60 seconds. Our evidence suggests that the mean time is actually less than 60 seconds.

d) According to both tests, there is evidence that the mean time required for rats to complete the maze is different than 60 seconds. The maze does not meet the "one-minute average" requirement. It should be noted that the test without the outlier is the appropriate test. The one slow rat made the mean time required seem much higher than it probably was.

Chapter 24 – Comparing Means

1. Learning math.

a) The margin of error of this confidence interval is (11.427 – 5.573)/2 = 2.927 points.

b) The margin of error for a 98% confidence interval would have been larger. The critical value of t^* is larger for higher confidence levels. We need a wider interval to increase the likelihood that we catch the true mean difference in test scores within our interval. In other words, greater confidence comes at the expense of precision.

c) We are 95% confident that the mean score for the CPMP math students will be between 5.573 and 11.427 points higher on this assessment than the mean score of the traditional students.

d) Since the entire interval is above 0, there is strong evidence that students who learn with CPMP will have higher mean scores is algebra than those in traditional programs.

3. CPMP, again.

a) H_0: The mean score of CPMP students is the same as the mean score of traditional students. $\left(\mu_C = \mu_T \text{ or } \mu_C - \mu_T = 0\right)$

H_A: The mean score of CPMP students is different from the mean score of traditional students. $\left(\mu_C \neq \mu_T \text{ or } \mu_C - \mu_T \neq 0\right)$

b) **Independent groups assumption:** Scores of students from different classes should be independent.
Randomization condition: Although not specifically stated, classes in this experiment were probably randomly assigned to either CPMP or traditional curricula.
10% condition: 312 and 265 are less than 10% of all students.
Nearly Normal condition: We don't have the actual data, so we can't check the distribution of the sample. However, the samples are large. The Central Limit Theorem allows us to proceed.

Since the conditions are satisfied, we can use a two-sample t-test with 583 degrees of freedom (from the computer).

c) If the mean scores for the CPMP and traditional students are really equal, there is less than a 1 in 10,000 chance of seeing a difference as large or larger than the observed difference just from natural sampling variation.

d) Since the P-value < 0.0001, reject the null hypothesis. There is strong evidence that the CPMP students have a different mean score than the traditional students. The evidence suggests that the CPMP students have a higher mean score.

5. **Commuting.**

   a) **Independent groups assumption:** Since the choice of route was determined at random, the commuting times for Route A are independent of the commuting times for Route B.
   Randomization condition: The man randomly determined which route he would travel on each day.
   10% condition: 20 days on each route represents less than 10% of all possible days.
   Nearly Normal condition: The histograms of travel times for the routes are roughly unimodal and symmetric. (Given)

   Since the conditions are satisfied, it is appropriate to model the sampling distribution of the difference in means with a Student's t-model, with 33.1 degrees of freedom (from the approximation formula). We will construct a two-sample t-interval, with 95% confidence.

   $$(\bar{y}_B - \bar{y}_A) \pm t^*_{df} \sqrt{\frac{s_B^2}{n_B} + \frac{s_A^2}{n_A}} = (43 - 40) \pm t^*_{33.1} \sqrt{\frac{2^2}{20} + \frac{3^2}{20}} \approx (1.36, 4.64)$$

   We are 95% confident that Route B has a mean commuting time between 1.36 and 4.64 minutes longer than the mean commuting time of Route A.

   b) Since 5 minutes is beyond the high end of the interval, there is no evidence that the Route B is an average of 5 minutes longer than Route A. It appears that the old-timer may be exaggerating the average difference in commuting time.

7. **Cereal.**

   Independent groups assumption: The percentage of sugar in the children's cereals is unrelated to the percentage of sugar in adult's cereals.
   Randomization condition: It is reasonable to assume that the cereals are representative of all children's cereals and adult cereals, in regard to sugar content.
   10% condition: 19 and 28 are less than 10% of all cereals.
   Nearly Normal condition: The histogram of adult cereal sugar content is skewed to the right, but the sample sizes are of reasonable size. The Central Limit Theorem allows us to proceed.

   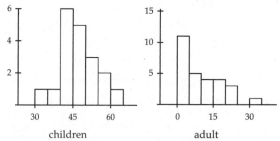

   Since the conditions are satisfied, it is appropriate to model the sampling distribution of the difference in means with a Student's t-model, with 42 degrees of freedom (from the approximation formula). We will construct a two-sample t-interval, with 95% confidence.

   $$(\bar{y}_C - \bar{y}_A) \pm t^*_{df} \sqrt{\frac{s_C^2}{n_C} + \frac{s_A^2}{n_A}} = (46.8 - 10.1536) \pm t^*_{42} \sqrt{\frac{6.41838^2}{19} + \frac{7.61239^2}{28}} \approx (32.49, 40.80)$$

   We are 95% confident that children's cereals have a mean sugar content that is between 32.49% and 40.80% higher than the mean sugar content of adult cereals.

9. Reading.

H_0: The mean reading comprehension score of students who learn by the new method is the same as the mean score of students who learn by traditional methods.
$$\left(\mu_N = \mu_T \ \text{ or } \ \mu_N - \mu_T = 0\right)$$

H_A: The mean reading comprehension score of students who learn by the new method is greater than the mean score of students who learn by traditional methods.
$$\left(\mu_N > \mu_T \ \text{ or } \ \mu_N - \mu_T > 0\right)$$

Independent groups assumption: Student scores in one group should not have an impact on the scores of students in the other group.
Randomization condition: Students were randomly assigned to classes.
10% condition: 18 and 20 are less than 10% of all students.
Nearly Normal condition: The histograms of the scores are unimodal and symmetric.

Since the conditions are satisfied, it is appropriate to model the sampling distribution of the difference in means with a Student's *t*-model, with 33 degrees of freedom (from the approximation formula). We will perform a two-sample *t*-test. We know:

$$\bar{y}_N = 51.7222 \qquad \bar{y}_T = 41.8$$
$$s_N = 11.7062 \qquad s_T = 17.4495$$
$$n_N = 18 \qquad n_T = 20$$

The sampling distribution model has mean 0, with standard error:

$$SE(\bar{y}_N - \bar{y}_T) = \sqrt{\frac{11.7062^2}{18} + \frac{17.4495^2}{20}} \approx 4.779 .$$

The observed difference between the mean scores is $51.7222 - 41.8 \approx 9.922$.

Since the *P*-value = 0.0228 is low, we reject the null hypothesis. There is evidence that the students taught using the new activities have a higher mean score on the reading comprehension test than the students taught using traditional methods.

$$t = \frac{(\bar{y}_N - \bar{y}_T) - (0)}{SE(\bar{y}_N - \bar{y}_T)}$$
$$t \approx \frac{9.922}{4.779}$$
$$t \approx 2.076$$

$P = 0.0228$

t_{33}

9.92

$t = 2.076$

11. Hurricanes.

1944-1969 1970-2000   1944-1969   1970-2000

The histograms of the number of hurricanes per year for the two time periods are both skewed, but that's not a problem. The sample sizes are large enough for the Central Limit Theorem to allow use to use our methods of inference. However, the outliers in the data prevent us from using inference. The Central Limit Theorem can't help us out here. It's probably best not to proceed with a test.

13. Baseball.

a)

American   National   American   National

The distributions are Nearly Normal, except for an outlier in the National League.

b) $\bar{y} \pm t^*_{n-1}\left(\dfrac{s}{\sqrt{n}}\right) = 9.80714 \pm t^*_{13}\left(\dfrac{0.805237}{\sqrt{14}}\right) \approx (9.34, 10.27)$

We are 95% confident that the mean number of runs scored per game in American League stadiums is between 9.34 and 10.27.

c) The data for the National League stadiums contains an outlier. In order to use the methods of inference, the outlier would have to be removed, and mentioned separately.

d) The average of 14 runs scored per game in Coors Field is unusual. It is almost 2 IQRs above quartile three, more than enough to classify as an outlier. Looking at the boxplot, it just seems out of place.

e) If you attempt to use two confidence intervals to assess a difference in means, you are actually adding standard deviations. But it's the variances that add, not the standard deviations. The two-sample difference of means procedure takes this into account.

15. Double header.

a) $(\bar{y}_A - \bar{y}_N) \pm t^*_{df} \sqrt{\dfrac{s^2_A}{n_A} + \dfrac{s^2_N}{n_N}} = (9.80714 - 9.58125) \pm t^*_{23} \sqrt{\dfrac{0.805237^2}{14} + \dfrac{1.54929^2}{16}} \approx (-0.69, 1.14)$

b) We are 95% confident that the mean number of runs scored in American League stadiums is between 0.69 runs lower and 1.14 runs higher than the mean number of runs scored in National League stadiums.

c) Since the interval contains 0, there is no evidence of a difference in the mean number of runs scored per game in the stadiums of the two leagues.

d) $(\bar{y}_A - \bar{y}_N) \pm t^*_{df} \sqrt{\dfrac{s^2_A}{n_A} + \dfrac{s^2_N}{n_N}} = (9.80714 - 9.28667) \pm t^*_{26} \sqrt{\dfrac{0.805237^2}{14} + \dfrac{1.04120^2}{15}} \approx (-0.19, 1.23)$

With the mean number of runs scored at Coors Field removed, we are 95% confident that the mean number of runs scored in American League stadiums is between 0.19 runs lower and 1.23 runs higher than the mean number of runs scored in National League stadiums. Since 0 is still in the interval, there is no evidence of a difference in the mean number of runs scored per game in the stadiums of the two leagues.

17. Job satisfaction.

A two-sample *t*-procedure is not appropriate for these data, because the two groups are not independent. They are before and after satisfaction scores for the same workers. Workers that have high levels of job satisfaction before the exercise program is implemented may tend to have higher levels of job satisfaction than other workers after the program as well.

19. Sex and violence.

a) Since the *P*-value = 0.136 is high, we fail to reject the null hypothesis. There is no evidence of a difference in the mean number of brands recalled by viewers of sexual content and viewers of violent content.

b) H_0: The mean number of brands recalled is the same for viewers of sexual content and viewers of neutral content. $(\mu_S = \mu_N \text{ or } \mu_S - \mu_N = 0)$

H_A: The mean number of brands recalled is different for viewers of sexual content and viewers of neutral content. $(\mu_S \neq \mu_N \text{ or } \mu_S - \mu_N \neq 0)$

Independent groups assumption: Recall of one group should not affect recall of another.
Randomization condition: Subjects were randomly assigned to groups.
10% condition: 108 and 108 are less than 10% of all TV viewers.
Nearly Normal condition: The samples are large.

Since the conditions are satisfied, it is appropriate to model the sampling distribution of the difference in means with a Student's *t*-model, with 214 degrees of freedom (from the approximation formula). We will perform a two-sample *t*-test.

The sampling distribution model has mean 0, with standard error:

$$SE(\bar{y}_S - \bar{y}_N) = \sqrt{\dfrac{1.76^2}{108} + \dfrac{1.77^2}{108}} \approx 0.24 \, .$$

The observed difference between the mean scores is 1.71 – 3.17 = – 1.46.

Since the *P*-value = 5.5×10^{-9} is low, we reject the null hypothesis. There is strong evidence that the mean number of brand names recalled is different for viewers of sexual content and viewers of neutral content. The evidence suggests that viewers of neutral ads remember more brand names on average than viewers of sexual content.

$$t = \frac{(\bar{y}_S - \bar{y}_N) - (0)}{SE(\bar{y}_S - \bar{y}_N)}$$

$$t \approx \frac{-1.46}{0.24}$$

$$t \approx -6.08$$

21. Sex and violence II.

a) H_0: The mean number of brands recalled is the same for viewers of violent content and viewers of neutral content. $(\mu_V = \mu_N$ or $\mu_V - \mu_N = 0)$

H_A: The mean number of brands recalled is different for viewers of violent content and viewers of neutral content. $(\mu_V \neq \mu_N$ or $\mu_V - \mu_N \neq 0)$

Independent groups assumption: Recall of one group should not affect recall of another.
Randomization condition: Subjects were randomly assigned to groups.
10% condition: 101 and 103 are less than 10% of all TV viewers.
Nearly Normal condition: The samples are large.

Since the conditions are satisfied, it is appropriate to model the sampling distribution of the difference in means with a Student's *t*-model, with 201.96 degrees of freedom (from the approximation formula). We will perform a two-sample *t*-test.

The sampling distribution model has mean 0, with standard error:

$$SE(\bar{y}_V - \bar{y}_N) = \sqrt{\frac{1.61^2}{101} + \frac{1.62^2}{103}} \approx 0.226.$$

The observed difference between the mean scores is 3.02 – 4.65 = – 1.63.

$$t = \frac{(\bar{y}_V - \bar{y}_N) - (0)}{SE(\bar{y}_V - \bar{y}_N)}$$

$$t \approx \frac{-1.63}{0.226}$$

$$t \approx -7.21$$

Since the *P*-value = 1.1×10^{-11} is low, we reject the null hypothesis. There is strong evidence that the mean number of brand names recalled is different for viewers of violent content and viewers of neutral content. The evidence suggests that viewers of neutral ads remember more brand names on average than viewers of violent content.

b) $(\bar{y}_N - \bar{y}_S) \pm t_{df}^* \sqrt{\frac{s_N^2}{n_N} + \frac{s_S^2}{n_S}} = (4.65 - 2.72) \pm t_{204.8}^* \sqrt{\frac{1.62^2}{103} + \frac{1.85^2}{106}} \approx (1.456, 2.404)$

We are 95% confident that the mean number of brand names recalled 24 hours later is between 1.46 and 2.40 higher for viewers of shows with neutral content than for viewers of shows with sexual content.

23. Lower scores?

a) Assuming that the conditions for inference were met by the NAEP, a 95% confidence interval for the difference in mean score is:

$$(\bar{y}_{1996} - \bar{y}_{2000}) \pm t_{df}^* \sqrt{\frac{s_{1996}^2}{n_{1996}} + \frac{s_{2000}^2}{n_{2000}}} = (150 - 147) \pm 1.960(1.22) \approx (0.61, 5.39)$$

Since the samples sizes are very large, it should be safe to use $z^* = 1.960$ for the critical value of t. We are 95% confident that the mean score in 2000 was between 0.61 and 5.39 points lower than the mean score in 1996. Since 0 is not contained in the interval, this provides evidence that the mean score has decreased from 1996 to 2000.

b) Both sample sizes are very large, which will make the standard errors of these samples very small. They are both likely to be very accurate. The difference in sample size shouldn't make you any more certain or any less certain.

However, these results are completely dependent upon whether or not the conditions for inference were met. If, by sampling more students, the NAEP sampled from a different population, then the two years are incomparable.

25. Statistics journals.

Independent groups assumption: These were articles submitted for publication in two different journals.
Randomization condition: It is necessary to assume that these articles are representative of all articles of this type with respect to publication delay.
10% condition: 288 and 209 are less than 10% of all articles submitted for publication to these journals.
Nearly Normal condition: The samples are large, so as long as there are no outliers in the data, it is okay to proceed.

Since the conditions are satisfied, it is appropriate to model the sampling distribution of the difference in means with a Student's t-model, with 338.30 degrees of freedom (from the approximation formula). We will construct a two-sample t-interval, with 90% confidence.

$$(\bar{y}_{Ap} - \bar{y}_{Am}) \pm t_{df}^* \sqrt{\frac{s_{Ap}^2}{n_{Ap}} + \frac{s_{Am}^2}{n_{Am}}} = (31 - 21) \pm t_{338.30}^* \sqrt{\frac{12^2}{209} + \frac{8^2}{288}} \approx (8.43, 11.57)$$

We are 90% confident that the mean number of months by which publication is delayed is between 8.43 and 11.57 months greater for *Applied Statistics* than it is for *The American Statistician*.

27. Mozart.

a) H_0: The mean memory test score is the same for those who listen to rap as it is for those who listen to no music. $\left(\mu_R = \mu_N \ \text{or} \ \mu_R - \mu_N = 0\right)$

H_A: The mean memory test score is lower for those who listen to rap than it is for those who listen to no music. $\left(\mu_R < \mu_N \ \text{or} \ \mu_R - \mu_N < 0\right)$

Independent groups assumption: The groups are not related in regards to memory score.
Randomization condition: Subjects were randomly assigned to groups.
10% condition: 29 and 13 are less than 10% of all people.
Nearly Normal condition: We don't have the actual data. We will assume that the distributions of the populations of memory test scores are Normal.

Since the conditions are satisfied, it is appropriate to model the sampling distribution of the difference in means with a Student's t-model, with 20.00 degrees of freedom (from the approximation formula). We will perform a two-sample t-test.

The sampling distribution model has mean 0, with standard error:

$$SE(\bar{y}_R - \bar{y}_N) = \sqrt{\frac{3.99^2}{29} + \frac{4.73^2}{13}} \approx 1.5066.$$

The observed difference between the mean number of objects remembered is 10.72 – 12.77 = – 2.05.

$t = \dfrac{(\bar{y}_R - \bar{y}_N) - (0)}{SE(\bar{y}_R - \bar{y}_N)}$

$t \approx \dfrac{-2.05}{1.5066}$

$t \approx -1.36$

$P = 0.0944$

Since the P-value = 0.0944 is high, we fail to reject the null hypothesis. There is little evidence that the mean number of objects remembered by those who listen to rap is lower than the mean number of objects remembered by those who listen to no music.

b) We did not conclude that there was a difference in the number of items remembered.

Chapter 25 – Paired Samples and Blocks

1. **More eggs?**

   a) Randomly assign 50 hens to each of the two kinds of feed. Compare the mean egg production of the two groups at the end of one month.

   b) Randomly divide the 100 hens into two groups of 50 hens each. Feed the hens in the first group the regular feed for two weeks, then switch to the additive for 2 weeks. Feed the hens in the second group the additive for two weeks, then switch to the regular feed for two weeks. Subtract each hen's "regular" egg production from her "additive" egg production, and analyze the mean difference in egg production.

   c) The matched pairs design in part b is the stronger design. Hens vary in their egg production regardless of feed. This design controls for that variability by matching the hens with themselves.

3. **Women.**

   a) The paired *t*-test is appropriate. The labor force participation rate for two different years was paired by city.

   b) Since the *P*-value = 0.0244, there is evidence of a difference in the average labor force participation rate for women between 1968 and 1972. The evidence suggests an increase in the participation rate for women.

5. **Rain.**

   a) The two-sample *t*-test is appropriate for these data. The seeded and unseeded clouds are not paired in any way. They are independent.

   b) Since the *P*-value = 0.0538, there is some evidence that the mean rainfall from seeded clouds is greater than the mean rainfall from unseeded clouds.

7. **Temperatures.**

> **Paired data assumption:** The data are paired by city.
> **Randomization condition:** These cities might not be representative of all European cities, so be cautious in generalizing the results.
> **10% condition:** 12 cities are less than 10% of all European cities.
> **Normal population assumption:** The histogram of differences between January and July mean temperature is roughly unimodal and symmetric.

July-Jan

> Since the conditions are satisfied, the sampling distribution of the difference can be modeled with a Student's *t*-model with 12 – 1 = 11 degrees of freedom. We will find a paired *t*-interval, with 90% confidence.

$$\bar{d} \pm t_{n-1}^{*}\left(\frac{s_d}{\sqrt{n}}\right) = 36.8333 \pm t_{11}^{*}\left(\frac{8.66375}{\sqrt{12}}\right) \approx (32.3, \ 41.3)$$

> We are 90% confident that the average high temperature in European cities in July is an average of between 32.3° to 41.4° higher than in January.

9. **Push-ups.**

> **Independent groups assumption:** The group of boys is independent of the group of girls.
> **Randomization condition:** Assume that students are assigned to gym classes at random.
> **10% condition:** 12 boys and 12 girls are less than 10% of all kids.
> **Nearly Normal condition:** The histograms of the number of push-ups from each group are roughly unimodal and symmetric.

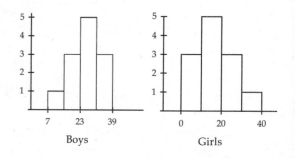

> Since the conditions are satisfied, it is appropriate to model the sampling distribution of the difference in means with a Student's *t*-model, with 21 degrees of freedom (from the approximation formula). We will construct a two-sample *t*-interval, with 90% confidence.

$$(\bar{y}_B - \bar{y}_G) \pm t_{df}^{*}\sqrt{\frac{s_B^2}{n_B} + \frac{s_G^2}{n_G}} = (23.8333 - 16.5000) \pm t_{21}^{*}\sqrt{\frac{7.20900^2}{12} + \frac{8.93919^2}{12}} \approx (1.6, \ 13.0)$$

> We are 90% confident that, at Gossett High, the mean number of push-ups that boys can do is between 1.6 and 13.0 more than the mean for the girls.

11. Job satisfaction.

a) Use a paired *t*-test.

Paired data assumption: The data are before and after job satisfaction rating for the same workers.
Randomization condition: The workers were randomly selected to participate.
10% condition: Assume that 10 workers are less than 10% of the workers at the company.
Nearly Normal conditon: The histogram of differences between before and after job satisfaction ratings is roughly unimodal and symmetric.

b) H₀: The mean difference in before and after job satisfaction scores is zero, and the exercise program is not effective at improving job satisfaction. $(\mu_d = 0)$

H_A: The mean difference in before and after job satisfaction scores is greater than zero, and the exercise program is effective at improving job satisfaction. $(\mu_d > 0)$

Since the conditions are satisfied, the sampling distribution of the difference can be modeled with a Student's *t*-model with 10 – 1 = 9 degrees of freedom, $t_9\left(0, \dfrac{7.47217}{\sqrt{10}}\right)$.

We will use a paired *t*-test, with $\bar{d} = 8.5$.

Since the *P*-value = 0.0029 is low, we reject the null hypothesis. There is evidence that the mean job satisfaction rating has increased since the implementation of the exercise program.

$$t = \frac{\bar{d} - 0}{\frac{s_d}{\sqrt{n}}}$$

$$t = \frac{8.5 - 0}{\frac{7.47217}{\sqrt{10}}}$$

$$t \approx 3.60$$

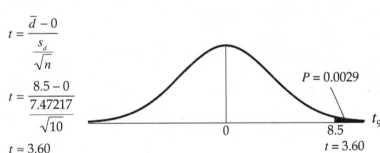

$P = 0.0029$

$t = 3.60$

13. Sleep.

a) H₀: The mean number of additional hours of sleep is zero. $(\mu_d = 0)$

H_A: The mean number of additional hours of sleep is greater than zero. $(\mu_d > 0)$

b) $P(t_9 > 3.680) = 0.0025$

c) If there is no gain of additional hours of sleep with the herb, the probability of observing a mean difference as large of larger than the one observed is 0.0025. (one-quarter percent)

d) Since the *P*-value = 0.0025 is low, we reject the null hypothesis. There is strong evidence that people taking the herb will get additional hours of sleep.

e) If we incorrectly reject the null hypothesis of no average gain in hours of sleep, we have committed a Type I error.

15. Yogurt.

H_0: The mean difference in calories between servings of strawberry and vanilla yogurt is zero. $(\mu_d = 0)$

H_A: The mean difference in calories between servings of strawberry and vanilla yogurt is different from zero. $(\mu_d \neq 0)$

Paired data assumption: The yogurt is paired by brand.
Randomization condition: Assume that these brands are representative of all brands.
10% condition: 12 brands of yogurt might not be less than 10% of all brands of yogurt. Proceed cautiously.
Normal population assumption: The histogram of differences in calorie content between strawberry and vanilla shows an outlier, Great Value. When the outlier is eliminated, the histogram of differences is roughly unimodal and symmetric.

With outlier   Without outlier

When Great Value yogurt is removed, the conditions are satisfied. The sampling distribution of the difference can be modeled with a Student's *t*-model with

11 – 1 = 10 degrees of freedom, $t_{10}\left(0, \dfrac{18.0907}{\sqrt{11}}\right)$.

We will use a paired *t*-test, with $\bar{d} \approx 4.54545$.

Since the *P*-value = 0.4241 is high, we fail to reject the null hypothesis. There is no evidence of a mean difference in calorie content between strawberry yogurt and vanilla yogurt.

$$t = \frac{\bar{d} - 0}{\dfrac{s_d}{\sqrt{n}}}$$

$$t = \frac{4.54545 - 0}{\dfrac{18.0907}{\sqrt{11}}}$$

$$t \approx 0.833$$

17. Braking.

a) **Randomization Condition:** These stops are probably representative of all such stops for this type of car, but not for all cars.
10% Condition: 10 stops are less than 10% of all possible stops.
Nearly Normal Condition: A histogram of the stopping distances is roughly unimodal and symmetric.

The stops in the sample had a mean stopping distance of 139.4 feet, and a standard deviation of 8.09938 feet. Since the conditions have been satisfied, construct a one-sample *t*-interval, with 10 – 1 = 9 degrees of freedom, at 95% confidence.

$$\bar{y} \pm t_{n-1}^{*}\left(\frac{s}{\sqrt{n}}\right) = 139.4 \pm t_{9}^{*}\left(\frac{8.09938}{\sqrt{10}}\right) \approx (133.6, 145.2)$$
We are 95% confident that the mean

dry pavement stopping distance for this type of car is between 133.6 and 145.2 feet.

b) **Independent Groups Assumption:** The wet pavement stops and dry pavement stops were made under different conditions and not paired in any way.
Randomization Condition: These stops are probably representative of all such stops for this type of car, but not for all cars.
10% Condition: 10 stops are less than 10% of all possible stops.
Nearly Normal Condition: The histogram of dry pavement stopping distances is roughly unimodal and symmetric (from part a), but the histogram of wet pavement stopping distances is a bit skewed. Since the Normal probability plot looks fairly straight, we will proceed.

Since the conditions are satisfied, it is appropriate to model the sampling distribution of the difference in means with a Student's *t*-model, with 13.8 degrees of freedom (from the approximation formula). We will construct a two-sample *t*-interval, with 95% confidence.

$$(\bar{y}_W - \bar{y}_D) \pm t^*_{df}\sqrt{\frac{s^2_W}{n_W} + \frac{s^2_D}{n_D}} = (202.4 - 139.4) \pm t^*_{13.8}\sqrt{\frac{15.07168^2}{10} + \frac{8.09938^2}{10}} \approx (51.4, 74.6)$$

We are 95% confident that the mean stopping distance on wet pavement is between 51.4 and 74.6 feet longer than the mean stopping distance on dry pavement.

19. Braking, test 2.

a) **Randomization Condition:** These cars are not a random sample, but are probably representative of all cars in terms of stopping distance.
10% Condition: These 10 cars are less than 10% of all cars.
Nearly Normal Condition: A histogram of the stopping distances is skewed to the right, but this may just be sampling variation from a Normal population. The "skew" is only a couple of stopping distances. We will proceed cautiously.

The cars in the sample had a mean stopping distance of 138.7 feet and a standard deviation of 9.66149 feet. Since the conditions have been satisfied, construct a one-sample *t*-interval, with 10 – 1 = 9 degrees of freedom, at 95% confidence.

$$\bar{y} \pm t^*_{n-1}\left(\frac{s}{\sqrt{n}}\right) = 138.7 \pm t^*_9\left(\frac{9.66149}{\sqrt{10}}\right) \approx (131.8, 145.6)$$

We are 95% confident that the mean dry pavement stopping distance for cars with this type of tires is between 131.8 and 145.6 feet. This estimate is based on an assumption that these cars are representative of all cars and that the population of stopping distances is Normal.

b) **Paired data assumption:** The data are paired by car.
Randomization condition: Assume that the cars are representative of all cars.
10% condition: The 10 cars tested are less than 10% of all cars.
Normal population assumption: The difference in stopping distance for car #4 is an outlier, at only 12 feet. After excluding this difference, the histogram of differences is unimodal and symmetric.

Since the conditions are satisfied, the sampling distribution of the difference can be modeled with a Student's t-model with $9 - 1 = 8$ degrees of freedom. We will find a paired t-interval, with 95% confidence.

$$\bar{d} \pm t^*_{n-1}\left(\frac{s_d}{\sqrt{n}}\right) = 55 \pm t^*_8\left(\frac{10.2103}{\sqrt{9}}\right) \approx (47.2, 62.8)$$

With car #4 removed, we are 95% confident that the mean increase in stopping distance on wet pavement is between 47.2 and 62.8 feet. (If you leave the outlier in, the interval is 38.8 to 62.6 feet, but you should remove it! This procedure is sensitive to the presence of outliers!)

21. Strikes.

a) Since 60% of 50 pitches is 30 pitches, the Little Leaguers would have to throw an average of more than 30 strikes in order to give support to the claim made by the advertisements.

H_0: The mean number of strikes thrown by Little Leaguers who have completed the training is 30. $(\mu_A = 30)$

H_A: The mean number of strikes thrown by Little Leaguers who have completed the training is greater than 30. $(\mu_A > 30)$

Randomization Condition: Assume that these players are representative of all Little League pitchers.
10% Condition: 20 pitchers are less than 10% of all Little League pitchers.
Nearly Normal Condition: The histogram of the number of strikes thrown after the training is roughly unimodal and symmetric.

The pitchers in the sample threw a mean of 33.15 strikes, with a standard deviation of 2.32322 strikes. Since the conditions for inference are satisfied, we can model the sampling distribution of the mean number of strikes thrown with a Student's t model, with $20 - 1 = 19$ degrees of freedom, $t_{19}\left(30, \frac{2.32322}{\sqrt{20}}\right)$.

We will perform a one-sample t-test.

Since the P-value $= 3.92 \times 10^{-6}$ is very low, we reject the null hypothesis. There is strong evidence that the mean number of strikes that Little Leaguers can throw after the training is more than 30. (This test says nothing about the effectiveness of the training; just that Little Leaguers can throw more than 60% strikes on average after completing the training. This might not be an improvement.)

$$t = \frac{\bar{y}_A - \mu_0}{\frac{s_A}{\sqrt{n_A}}}$$

$$t = \frac{33.15 - 30}{\frac{2.32322}{\sqrt{20}}}$$

$$t = 6.06$$

b) H_0: The mean difference in number of strikes thrown before and after the training is zero. $(\mu_d = 0)$

H_A: The mean difference in number of strikes thrown before and after the training is greater than zero. $(\mu_d > 0)$

Paired data assumption: The data are paired by pitcher.
Randomization condition: Assume that these players are representative of all Little League pitchers.
10% condition: 20 players are less than 10% of all players.
Normal population assumption: The histogram of differences is roughly unimodal and symmetric.

After-Before

Since the conditions are satisfied, the sampling distribution of the difference can be modeled with a Student's t-model with $20 - 1 = 19$ degrees of freedom, $t_{19}\left(0, \dfrac{3.32297}{\sqrt{19}}\right)$.

We will use a paired t-test, with $\bar{d} = 0.1$.

Since the P-value = 0.4472 is high, we fail to reject the null hypothesis. There is no evidence of a mean difference in number of strikes thrown before and after the training. The training does not appear to be effective.

$$t = \frac{\bar{d} - 0}{\frac{s_d}{\sqrt{n}}}$$

$$t = \frac{0.1 - 0}{\frac{3.32297}{\sqrt{20}}}$$

$$t \approx 0.135$$

$P = 0.4472$

t_{19}

$0 \quad 0.10$

$t = 0.135$

Review of Part VI – Learning About the World

1. Crawling.

   a) H_0: The mean age at which babies begin to crawl is the same whether the babies were born in January or July. $\left(\mu_{Jan} = \mu_{July} \text{ or } \mu_{Jan} - \mu_{July} = 0\right)$

   H_A: There is a difference in the mean age at which babies begin to crawl, depending on whether the babies were born in January or July. $\left(\mu_{Jan} \neq \mu_{July} \text{ or } \mu_{Jan} - \mu_{July} \neq 0\right)$

   Independent groups assumption: The groups of January and July babies are independent.
   Randomization condition: Although not specifically stated, we will assume that the babies are representative of all babies.
   10% condition: 32 and 21 are less than 10% of all babies.
   Nearly Normal condition: We don't have the actual data, so we can't check the distribution of the sample. However, the samples are fairly large. The Central Limit Theorem allows us to proceed.

   Since the conditions are satisfied, it is appropriate to model the sampling distribution of the difference in means with a Student's t-model, with 43.68 degrees of freedom (from the approximation formula).

   We will perform a two-sample t-test. The sampling distribution model has mean 0, with

   standard error: $SE(\bar{y}_{Jan} - \bar{y}_{July}) = \sqrt{\dfrac{7.08^2}{32} + \dfrac{6.91^2}{21}} \approx 1.9596$.

   The observed difference between the mean ages is 29.84 – 33.64 = – 3.8 weeks.

   Since the P-value = 0.0590 is fairly low, we reject the null hypothesis. There is some evidence that mean age at which babies crawl is different for January and June babies. June babies appear to crawl a bit earlier than July babies, on average. Since the evidence is not strong, we might want to do some more research into this claim.

   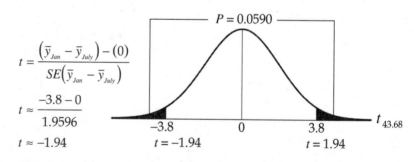

$$t = \frac{(\bar{y}_{Jan} - \bar{y}_{July}) - (0)}{SE(\bar{y}_{Jan} - \bar{y}_{July})}$$

$$t \approx \frac{-3.8 - 0}{1.9596}$$

$$t \approx -1.94$$

   b) H_0: The mean age at which babies begin to crawl is the same whether the babies were born in April or October. $\left(\mu_{Apr} = \mu_{Oct} \text{ or } \mu_{Apr} - \mu_{Oct} = 0\right)$

   H_A: There is a difference in the mean age at which babies begin to crawl, depending on whether the babies were born in April or October. $\left(\mu_{Apr} \neq \mu_{Oct} \text{ or } \mu_{Apr} - \mu_{Oct} \neq 0\right)$

   The conditions (with minor variations) were checked in part a.

Since the conditions are satisfied, it is appropriate to model the sampling distribution of the difference in means with a Student's *t*-model, with 59.40 degrees of freedom (from the approximation formula).

We will perform a two-sample *t*-test. The sampling distribution model has mean 0, with standard error: $SE(\bar{y}_{Apr} - \bar{y}_{Oct}) = \sqrt{\dfrac{6.21^2}{26} + \dfrac{7.29^2}{44}} \approx 1.6404$.

The observed difference between the mean ages is 31.84 – 33.35 = – 1.51 weeks.

Since the *P*-value = 0.3610 is high, we fail to reject the null hypothesis. There is no evidence that mean age at which babies crawl is different for April and October babies.

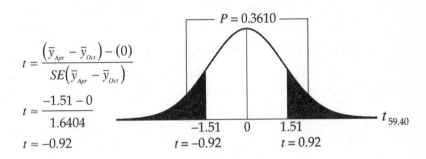

$$t = \frac{(\bar{y}_{Apr} - \bar{y}_{Oct}) - (0)}{SE(\bar{y}_{Apr} - \bar{y}_{Oct})}$$

$$t \approx \frac{-1.51 - 0}{1.6404}$$

$$t \approx -0.92$$

c) These results are not consistent with the researcher's claim. We have slight evidence in one test and no evidence in the other. The researcher will have to do better than this to convince us!

3. **Women.**

H_0 : The percentage of businesses in the area owned by women is 26%. ($p = 0.26$)
H_A : The percentage of businesses in the area owned by women is not 26%. ($p \neq 0.26$)

Random condition: This is a random sample of 410 businesses in the Denver area.
10% condition: The sample of 410 businesses is less than 10% of all businesses.
Success/Failure condition: $np = (410)(0.26) = 106.6$ and $nq = (410)(0.74) = 303.4$ are both greater than 10, so the sample is large enough.

The conditions have been satisfied, so a Normal model can be used to model the sampling distribution of the proportion, with $\mu_{\hat{p}} = p = 0.26$ and $\sigma(\hat{p}) = \sqrt{\dfrac{pq}{n}} = \sqrt{\dfrac{(0.26)(0.74)}{410}} \approx 0.02166$.

We can perform a one-proportion *z*-test. The observed proportion of businesses owned by women is $\hat{p} = \dfrac{115}{410} \approx 0.2805$.

Since the *P*-value = 0.3443 is high, we fail to reject the null hypothesis. There is no evidence that the proportion of businesses in the Denver area owned by women is any different than the national figure of 26%.

$$z = \frac{\hat{p} - p_0}{\sqrt{\dfrac{pq}{n}}}$$

$$z = \frac{0.2805 - 0.26}{\sqrt{\dfrac{(0.26)(0.74)}{410}}}$$

$$z \approx 0.946$$

5. Pottery.

> **Independent groups assumption:** The pottery samples are from two different sites.
> **Randomization condition:** It is reasonable to think that the pottery samples are representative of all pottery at that site with respect to aluminum oxide content.
> **10% condition:** The samples of 5 pieces are less than 10% of all pottery pieces.
> **Nearly Normal condition:** The histograms of aluminum oxide content are roughly unimodal and symmetric.

Ashley Rails   New Forest

Since the conditions are satisfied, it is appropriate to model the sampling distribution of the difference in means with a Student's t-model, with 7 degrees of freedom (from the approximation formula). We will construct a two-sample t-interval, with 95% confidence.

$$(\bar{y}_{AR} - \bar{y}_{NF}) \pm t^*_{df}\sqrt{\frac{s^2_{AR}}{n_{AR}} + \frac{s^2_{NF}}{n_{NF}}} = (17.32 - 18.18) \pm t^*_7\sqrt{\frac{1.65892^2}{5} + \frac{1.77539^2}{5}} \approx (-3.37, 1.65)$$

We are 95% confident that the difference in the mean percentage of aluminum oxide content of the pottery at the two sites is between –3.37% and 1.65%. Since 0 is in the interval, there is no evidence that the aluminum oxide content at the two sites is different. It would be reasonable for the archaeologists to think that the same ancient people inhabited the sites.

7. Gehrig.

> H_0: The proportion of ALS patients who were athletes is the same as the proportion of patients with other disorders who were athletes. $\left(p_{ALS} = p_{Other} \text{ or } p_{ALS} - p_{Other} = 0\right)$

> H_A: The proportion of ALS patients who were athletes is greater than the proportion of patients with other disorders who were athletes. $\left(p_{ALS} > p_{Other} \text{ or } p_{ALS} - p_{Other} > 0\right)$

> **Random condition:** This is NOT a random sample. We must assume that these patients are representative of all patients with neurological disorders.
> **10% condition:** 280 and 151 are both less than 10% of all patients with disorders.
> **Independent samples condition:** The groups are independent.
> **Success/Failure condition:** $n\hat{p}$(ALS) = (280)(0.38) = 106, $n\hat{q}$(ALS) = (280)(0.72) = 174, $n\hat{p}$(Other) = (151)(0.26) = 39, and $n\hat{q}$(Other) = (151)(0.74) = 112 are all greater than 10, so the samples are both large enough.

Since the conditions have been satisfied, we will model the sampling distribution of the difference in proportion with a Normal model with mean 0 and standard deviation estimated by

$$SE_{pooled}\left(\hat{p}_{ALS} - \hat{p}_{Other}\right) = \sqrt{\frac{\hat{p}_{pooled}\hat{q}_{pooled}}{n_{ALS}} + \frac{\hat{p}_{pooled}\hat{q}_{pooled}}{n_{Other}}} = \sqrt{\frac{(0.336)(0.664)}{280} + \frac{(0.336)(0.664)}{151}} \approx 0.0477.$$

The observed difference between the proportions is 0.38 – 0.26 = 0.12.

Since the *P*-value = 0.0060 is very low, we reject the null hypothesis. There is strong evidence that the proportion of ALS patients who are athletes is greater than the proportion of patients with other disorders who are athletes.

$$z = \frac{0.12 - 0}{0.0477}$$

$$z \approx 2.52$$

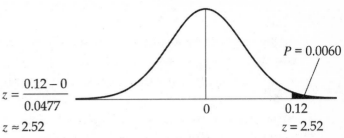

b) This was a retrospective observational study. In order to make the inference, we must assume that the patients studied are representative of all patients with neurological disorders.

9. Babies.

H_0: The mean weight of newborns in the U.S. is 7.41 pounds, the same as the mean weight of Australian babies. $(\mu = 7.41)$

H_A: The mean weight of newborns in the U.S. is not the same as the mean weight of Australian babies. $(\mu \neq 7.41)$

Randomization condition: Assume that the babies at this Missouri hospital are representative of all U.S. newborns. (Given)
10% condition: 112 newborns are less than 10% of all newborns.
Nearly Normal condition: We don't have the actual data, so we cannot look at a graphical display, but since the sample is large, it is safe to proceed.

The babies in the sample had a mean weight of 7.68 pounds and a standard deviation in weight of 1.31 pounds. Since the conditions for inference are satisfied, we can model the sampling distribution of the mean weight of U.S. newborns with a Student's *t* model, with

$112 - 1 = 111$ degrees of freedom, $t_{111}\left(7.41, \dfrac{1.31}{\sqrt{112}}\right)$.

We will perform a one-sample *t*-test.

Since the *P*-value = 0.0313 is low, we reject the null hypothesis. If we believe that the babies at this Missouri hospital are representative of all U.S. babies, there is evidence to suggest that the mean weight of U.S. babies is different than the mean weight of Australian babies. U.S. babies appear to weigh more on average.

$$t = \frac{\bar{y} - \mu_0}{SE(\bar{y})}$$

$$t = \frac{7.68 - 7.41}{\dfrac{1.31}{\sqrt{112}}}$$

$$t \approx 2.18$$

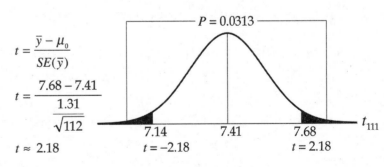

11. Feeding fish.

a) If there is no difference in the average fish sizes, the chance of observing a difference this large, or larger, just by natural sampling variation is 0.1%.

b) There is evidence that largemouth bass that are fed a natural diet are larger. The researchers would advise people who raise largemouth bass to feed them a natural diet.

c) If the advice is incorrect, the researchers have committed a Type I error.

13. Age.

a) **Independent groups assumption:** The group of patients with and without cardiac disease are not related in any way.
Randomization condition: Assume that these patients are representative of all people.
10% condition: 2397 patients without cardiac disease and 450 patients with cardiac disease are both less than 10% of all people.
Normal population assumption: We don't have the actual data, so we will assume that the population of ages of patients is Normal.

Since the conditions are satisfied, it is appropriate to model the sampling distribution of the difference in means with a Student's t-model, with 670 degrees of freedom (from the approximation formula). We will construct a two-sample t-interval, with 95% confidence.

$$(\bar{y}_{Card} - \bar{y}_{None}) \pm t_{df}^* \sqrt{\frac{s_{Card}^2}{n_{Card}} + \frac{s_{None}^2}{n_{None}}} = (74.0 - 69.8) \pm t_{670}^* \sqrt{\frac{7.9^2}{450} + \frac{8.7^2}{2397}} \approx (3.39, 5.01)$$

We are 95% confident that the mean age of patients with cardiac disease is between 3.39 and 5.01 years higher than the mean age of patients without cardiac disease.

b) Older patients are at greater risk for a variety of health problems. If an older patient does not survive a heart attack, the researchers will not know to what extent depression was involved, because there will be a variety of other possible variables influencing the death rate. Additionally, older patients may be more (or less) likely to be depressed than younger ones.

15. Computer use.

a) It is unlikely that an equal number of boys and girls were contacted strictly by chance. It is likely that this was a stratified random sample, stratified by gender.

b) **Randomization condition:** The teens were selected at random.
10% condition: 620 boys and 620 girls are both less than 10% of all teens.
Independent groups assumption: The groups of boys and girls are not paired or otherwise related in any way.
Success/Failure condition: $n\hat{p}$ (boys) = (620)(0.77) = 477, $n\hat{q}$ (boys) = (620)(0.23) = 143, $n\hat{p}$ (girls) = (620)(0.65) = 403, and $n\hat{q}$ (girls) = (620)(0.35) = 217 are all greater than 10, so the samples are both large enough.

Since the conditions have been satisfied, we will find a two-proportion z-interval.

$$(\hat{p}_B - \hat{p}_G) \pm z^* \sqrt{\frac{\hat{p}_B \hat{q}_B}{n_B} + \frac{\hat{p}_G \hat{q}_G}{n_G}} = (0.77 - 0.65) \pm 1.960 \sqrt{\frac{(0.77)(0.33)}{620} + \frac{(0.65)(0.35)}{620}} = (0.070, 0.170)$$

We are 95% confident that the proportion of computer gamers is between 7.0% and 17.0% higher for boys than for girls.

c) Since the interval lies entirely above 0, there is evidence that a greater percentage of boys play computer games than girls.

17. Hearing.

Paired data assumption: The data are paired by subject.
Randomization condition: The order of the tapes was randomized.
10% condition: We are testing the tapes, not the people, so this condition does not need to be checked.
Normal population assumption: The histogram of differences between List A and List B is roughly unimodal and symmetric.

Since the conditions are satisfied, the sampling distribution of the difference can be modeled with a Student's t-model with $24 - 1 = 23$ degrees of freedom. We will find a paired t-interval, with 95% confidence.

$$\bar{d} \pm t_{n-1}^{*}\left(\frac{s_d}{\sqrt{n}}\right) = -0.\overline{3} \pm t_{23}^{*}\left(\frac{8.12225}{\sqrt{24}}\right) \approx (-3.76, 3.10)$$

We are 95% confident that the mean difference in the number of words a person might misunderstand using these two lists is between – 3.76 and 3.10 words. Since 0 is contained in the interval, there is no evidence to suggest that that the two lists are different for the purposes of the hearing test when there is background noise. It is reasonable to think that the two lists are still equivalent.

19. Newspapers.

a) An examination of a graphical display reveals Spain, Portugal, and Italy to be outliers. They are all Mediterranean countries, and all have a significantly higher percentage of men than women reading a newspaper daily.

Men - Women

b) H$_0$: The mean difference in the percentage of men and women who read a daily newspaper in these countries is zero. $\left(\mu_d = 0\right)$

H$_A$: The mean difference in the percentage of men and women who read a daily newspaper in these countries is greater than zero. $\left(\mu_d > 0\right)$

Men - Women

Paired data assumption: The data are paired by country.
Randomization condition: Samples in each country were random.
10% condition: The 1000 respondents in each country are less than 10% of the people in these countries.
Nearly Normal condition: With three outliers removed, the distribution of differences is roughly unimodal and symmetric.

Since the conditions are satisfied, the sampling distribution of the difference can be modeled with a Student's t-model with $11 - 1 = 10$ degrees of freedom, $t_{10}\left(0, \frac{2.83668}{\sqrt{11}}\right)$.

We will use a paired t-test, (Men – Women) with $\bar{d} = 4.75455$.

Since the *P*-value = 0.0001 is very low, we reject the null hypothesis. There is strong evidence that the mean difference is greater than zero. The percentage of men in these countries who read the paper daily appears to be greater than the percentage of women who do so.

$$t = \frac{\bar{d} - 0}{\frac{s_d}{\sqrt{11}}}$$

$$t \approx \frac{4.75455 - 0}{\frac{2.83668}{\sqrt{11}}}$$

$$t \approx 5.56$$

21. Wall Street.

Random condition: 1002 American adults were randomly selected.
10% condition: 1002 is less than 10% of all American adults.
Success/Failure condition: $n\hat{p} = (1002)(0.60) = 601$ and $n\hat{q} = (1002)(0.40) = 401$ are both greater than 10, so the sample is large enough.

Since the conditions are met, we can use a one-proportion *z*-interval to estimate the percentage of American adults who agree with the statement.

$$\hat{p} \pm z^* \sqrt{\frac{\hat{p}\hat{q}}{n}} = (0.60) \pm 1.960 \sqrt{\frac{(0.60)(0.40)}{1002}} \approx (57.0\%, 63.0\%)$$

We are 95% confident that between 57.0% and 63.0% of American adults would agree with the statement about Wall Street.

23. Legionnaires' disease.

Paired data assumption: The data are paired by room.
Randomization condition: We will assume that these rooms are representative of all rooms at the hotel.
10% condition: Assume that 8 rooms are less than 10% of the rooms. The hotel must have more than 80 rooms.
Nearly Normal condition: The histogram of differences between before and after measurements is roughly unimodal and symmetric.

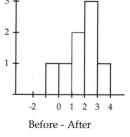

Before - After

Since the conditions are satisfied, the sampling distribution of the difference can be modeled with a Student's *t*-model with 8 – 1 = 7 degrees of freedom. We will find a paired *t*-interval, with 95% confidence.

$$\bar{d} \pm t^*_{n-1}\left(\frac{s_d}{\sqrt{n}}\right) = 1.6125 \pm t^*_7\left(\frac{1.23801}{\sqrt{8}}\right) \approx (0.58, 2.65)$$

We are 95% confident that the mean difference in the bacteria counts is between 0.58 and 2.65 colonies per cubic foot of air. Since the entire interval is above 0, there is evidence that the new air-conditioning system was effective in reducing average bacteria counts.

25. Bipolar kids.

a) **Random condition:** Assume that the 89 children are representative of all children with bipolar disorder.
10% condition: 89 is less than 10% of all children with bipolar disorder.
Success/Failure condition: $n\hat{p} = 26$ and $n\hat{q} = 63$ are both greater than 10, so the sample is large enough.

Since the conditions are met, we can use a one-proportion z-interval to estimate the percentage of children with bipolar disorder who might be helped by this treatment.

$$\hat{p} \pm z^* \sqrt{\frac{\hat{p}\hat{q}}{n}} = \left(\frac{26}{89}\right) \pm 1.960 \sqrt{\frac{\left(\frac{26}{89}\right)\left(\frac{63}{89}\right)}{89}} \approx (19.77\%, 38.66\%)$$

We are 95% confident that between 19.77% and 38.66% of children with bipolar disorder will be helped with medication and psychotherapy.

b)

$$ME = z^* \sqrt{\frac{\hat{p}\hat{q}}{n}}$$

$$0.06 = 1.960 \sqrt{\frac{\left(\frac{26}{89}\right)\left(\frac{63}{89}\right)}{n}}$$

$$n = \frac{(1.960)^2 \left(\frac{26}{89}\right)\left(\frac{63}{89}\right)}{(0.06)^2}$$

$$n \approx 221 \text{ children}$$

In order to estimate the proportion of children helped with medication and psychotherapy to within 6% with 95% confidence, we would need a sample of at least 221 children. All decimals in the final answer must be rounded up, to the next person.
(For a more cautious answer, let $\hat{p} = \hat{q} = 0.5$. This method results in a required sample of 267 children.)

27. Bread.

a) Since the histogram shows that the distribution of the number of loaves sold per day is skewed strongly to the right, we can't use the Normal model to estimate the number of loaves sold on the busiest 10% of days.

b) **Randomization condition:** Assume that these days are representative of all days.
10% condition: 100 days are less than 10% of all days.
Nearly Normal condition: The histogram is skewed strongly to the right. However, since the sample size is large, the Central Limit Theorem guarantees that the distribution of averages will be approximately Normal.

The days in the sample had a mean of 103 loaves sold and a standard deviation of 9 loaves sold. Since the conditions are satisfied, the sampling distribution of the mean can be modeled by a Student's t- model, with $103 - 1 = 103$ degrees of freedom. We will use a one-sample t-interval with 95% confidence for the mean number of loaves sold.
(By hand, use $t^*_{50} \approx 2.403$ from the table.)

c) $\bar{y} \pm t^*_{n-1}\left(\frac{s}{\sqrt{n}}\right) = 103 \pm t^*_{102}\left(\frac{9}{\sqrt{103}}\right) \approx (101.2, 104.8)$

We are 95% confident that the mean number of loaves sold per day at the Clarksburg Bakery is between 101.2 and 104.8.

d) We know that in order to cut the margin of error in half, we need to a sample four times as large. If we allow a margin of error that is twice as wide, that would require a sample only one-fourth the size. In this case, our original sample is 100 loaves; so 25 loaves would be a sufficient number to estimate the mean with a margin of error twice as wide.

e) Since the interval is completely above 100 loaves, there is strong evidence that the estimate was incorrect. The evidence suggests that the mean number of loaves sold per day is greater than 100. This difference is statistically significant, but may not be practically significant. It seems like the owners made a pretty good estimate!

29. Insulin and diet.

a) H_0: People with high dairy consumption have IRS at the same rate as those with low dairy consumption. $\left(p_{High} = p_{Low} \ \text{ or } \ p_{High} - p_{Low} = 0\right)$

H_A: People with high dairy consumption have IRS at a different rate than those with low dairy consumption. $\left(p_{High} \neq p_{Low} \ \text{ or } \ p_{High} - p_{Low} \neq 0\right)$

Random condition: Assume that the people studied are representative of all people.
10% condition: 102 and 190 are both less than 10% of all people.
Independent samples condition: The two groups are not related.
Success/Failure condition: $n\hat{p}$ (high) = 24, $n\hat{q}$ (high) = 78, $n\hat{p}$ (low) = 85, and $n\hat{q}$ (low) = 105 are all greater than 10, so the samples are both large enough.

Since the conditions have been satisfied, we will model the sampling distribution of the difference in proportion with a Normal model with mean 0 and standard deviation estimated by

$$SE_{pooled}\left(\hat{p}_{High} - \hat{p}_{Low}\right) = \sqrt{\frac{\hat{p}_{pooled}\hat{q}_{pooled}}{n_{High}} + \frac{\hat{p}_{pooled}\hat{q}_{pooled}}{n_{Low}}} = \sqrt{\frac{(0.373)(0.627)}{102} + \frac{(0.373)(0.627)}{190}} \approx 0.05936.$$

The observed difference between the proportions is 0.2352 – 0.4474 = – 0.2122.

Since the P-value = 0.0004 is very low, we reject the null hypothesis. There is strong evidence that the proportion of people with IRS is different for those who with high dairy consumption compared to those with low dairy consumption. People who consume dairy products more than 35 times per week appear less likely to have IRS than those who consume dairy products fewer than 10 times per week.

$$z = \frac{-0.2122 - 0}{0.05936}$$

$$z \approx -3.57$$

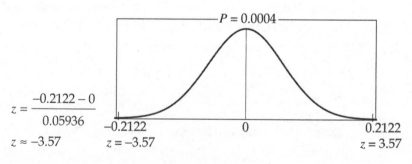

b) There is evidence of an association between the low consumption of dairy products and IRS, but that does not prove that dairy consumption influences the development of IRS. This is an observational study, and a controlled experiment is required to prove cause and effect.

31. Rainmakers?

Independent groups assumption: The two groups of clouds are independent.
Randomization condition: Researchers randomly assigned clouds to be seeded with silver iodide or not seeded.
10% condition: We are testing cloud seeding, not clouds themselves, so this condition doesn't need to be checked.
Nearly Normal condition: We don't have the actual data, so we can't look at the distributions, but the means of group are significantly higher than the medians. This is an indication that the distributions are skewed to the right, with possible outliers. The samples sizes of 26 each are fairly large, so it should be safe to proceed, but we should be careful making conclusions, since there may be outliers.

Since the conditions are satisfied, it is appropriate to model the sampling distribution of the difference in means with a Student's t-model, with 33.86 degrees of freedom (from the approximation formula). We will construct a two-sample t-interval, with 95% confidence.

$$(\bar{y}_S - \bar{y}_U) \pm t^*_{df} \sqrt{\frac{s_S^2}{n_S} + \frac{s_U^2}{n_U}} = (441.985 - 164.588) \pm t^*_{33.86} \sqrt{\frac{650.787^2}{26} + \frac{278.426^2}{26}} \approx (-4.76, 559.56)$$

We are 95% confident the mean amount of rainfall produced by seeded clouds is between 4.76 acre-feet less than and 559.56 acre-feet more than the mean amount of rainfall produced by unseeded clouds.

Since the interval contains 0, there is little evidence that the mean rainfall produced by seeded clouds is any different from the mean rainfall produced by unseeded clouds. However, we shouldn't place too much faith in this conclusion. It is based on a procedure that is sensitive to outliers, and there may have been outliers present.

33. Color or text?

a) By randomizing the order of the cards (shuffling), the researchers are attempting to avoid bias that may result from volunteers remembering the order of the cards from the first task. Although mentioned, hopefully the researchers are randomizing the order in which the tasks are performed. For example, if each volunteer performs the color task first, they may all do better on the text task, simply because they have had some practice in memorizing cards. By randomizing the order, that bias is controlled.

b) H₀: The mean difference between color and word scores is zero. $(\mu_d = 0)$

   Hₐ: The mean difference between color and word scores is not zero. $(\mu_d \neq 0)$

c) **Paired data assumption:** The data are paired by volunteer.
   Randomization condition: Hopefully, the volunteers were randomized with respect to the order in which they performed the tasks, and the cards were shuffled between tasks.
   10% condition: We are testing the format of the task (color/word), not the people, so we don't need to check this condition.
   Nearly Normal condition: The histogram of the differences is roughly unimodal and symmetric.

Color - Word

d) Since the conditions are satisfied, the sampling distribution of the difference can be modeled with a Student's *t*-model with 32 – 1 = 31 degrees of freedom, $t_{31}\left(0, \dfrac{2.50161}{\sqrt{32}}\right)$.

We will use a paired *t*-test, (color – word) with $\bar{d} = -0.75$.

e) Since the *P*-value = 0.0999 is high, we fail to reject the null hypothesis. There is no evidence that the mean difference in score is different from zero. It is reasonable to think that neither color nor word dominates perception. However, if the order in which volunteers took the test was not randomized, making conclusions from this test may be risky.

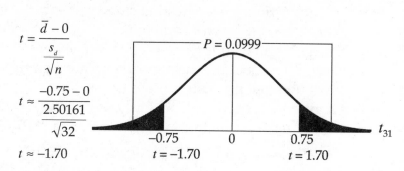

$$t = \frac{\bar{d} - 0}{\frac{s_d}{\sqrt{n}}}$$

$$t \approx \frac{-0.75 - 0}{\frac{2.50161}{\sqrt{32}}}$$

$$t \approx -1.70$$

35. Batteries.

a) Different samples have different means. Since this is a fairly small sample, the difference may be due to natural sampling variation. Also, we have no idea how to quantify "a lot less" with out considering the variation as measured by the standard deviation.

b) H_0: The mean life of a battery is 100 hours. $(\mu = 100)$
H_A: The mean life of a battery is less than 100 hours. $(\mu < 100)$

c) **Randomization condition:** It is reasonable to think that these 16 batteries are representative of all batteries of this type
10% condition: 16 batteries are less than 10% of all batteries.
Normal population assumption: Since we don't have the actual data, we can't check a graphical display, and the sample is not large. Assume that the population of battery lifetimes is Normal.

d) The batteries in the sample had a mean life of 97 hours and a standard deviation of 12 hours. Since the conditions for inference are satisfied, we can model the sampling distribution of the mean battery life with a Student's *t* model, with 16 – 1 = 15 degrees of freedom, $t_{15}\left(100, \dfrac{12}{\sqrt{16}}\right)$, or $t_{15}(100,3)$

We will perform a one-sample *t*-test.

Since the *P*-value = 0.1666 is greater than $\alpha = 0.05$, we fail to reject the null hypothesis. There is no evidence to suggest that the mean battery life is less than 100 hours.

$$t = \frac{\bar{y} - \mu_0}{SE(\bar{y})}$$

$$t = \frac{97 - 100}{\frac{12}{\sqrt{16}}}$$

$$t = -1$$

e) If the mean life of the company's batteries is only 98 hours, then the mean life is less than 100, and the null hypothesis is false. We failed to reject a false null hypothesis, making a Type II error.

Chapter 26 – Comparing Counts

1. **Dice.**

   a) If the die were fair, you'd expect each face to show 10 times.

   b) Use a chi-square test for goodness-of-fit. We are comparing the distribution of a single variable (outcome of a die roll) to an expected distribution.

   c) H_0: The die is fair. (All faces have the same probability of coming up.)

   H_A: The die is not fair. (Some faces are more or less likely to come up than others.)

   d) **Counted data condition:** We are counting the number of times each face comes up.
   Randomization condition: Die rolls are random and independent of each other.
   Expected cell frequency condition: We expect each face to come up 10 times, and 10 is greater than 5.

   e) Under these conditions, the sampling distribution of the test statistic is χ^2 on $6 - 1 = 5$ degrees of freedom. We will use a chi-square goodness-of-fit test.

Face	Observed	Expected	Residual = $(Obs - Exp)$	$(Obs - Exp)^2$	Component = $\dfrac{(Obs - Exp)^2}{Exp}$
1	11	10	1	1	0.1
2	7	10	– 3	9	0.9
3	9	10	– 1	1	0.1
4	15	10	5	25	2.5
5	12	10	2	4	0.4
6	6	10	– 4	16	1.6

$$\sum = 5.6$$

Since the *P*-value = 0.3471 is high, we fail to reject the null hypothesis. There is no evidence that the die is unfair.

3. **Nuts.**

   a) The weights of the nuts are quantitative. Chi-square goodness-of-fit requires counts.

   b) In order to use a chi-square test, you could count the number of each type of nut. However, it's not clear whether the company's claim was a percentage by number or a percentage by weight.

5. NYPD and race.

H_0: The distribution of ethnicities in the police department represents the distribution of ethnicities of the youth of New York City.

H_A: The distribution of ethnicities in the police department does not represent the distribution of ethnicities of the youth of New York City.

Counted data condition: The percentages reported must be converted to counts.

Randomization condition: Assume that the current NYPD is representative of recent departments with respect to ethnicity.

Expected cell frequency condition: The expected counts are all much greater than 5.

Ethnicity	Observed	Expected
White	16965	7644.852
Black	3796	7383.042
Latino	5001	8247.015
Asian	367	2382.471
Other	52	523.620

(Note: The observed counts should be whole numbers. They are actual policemen. The expected counts may be decimals, since they behave like averages.)

Under these conditions, the sampling distribution of the test statistic is χ^2 on $5 - 1 = 4$ degrees of freedom. We will use a chi-square goodness-of-fit test.

$$\chi^2 = \sum_{all\ cells} \frac{(Obs - Exp)^2}{Exp} = \frac{(16965 - 7644.852)^2}{7644.852} + \frac{(3796 - 7383.042)^2}{7383.042} + \frac{(5001 - 8247.015)^2}{8247.015}$$

$$+ \frac{(367 - 2382.471)^2}{2382.471} + \frac{(52 - 523.620)^2}{523.620} \approx 16,500$$

With χ^2 of over 16,500, on 4 degrees of freedom, the *P*-value is essentially 0, so we reject the null hypothesis. There is strong evidence that the distribution of ethnicities of NYPD officers does not represent the distribution of ethnicities of the youth of New York City. Specifically, the proportion of white officers is much higher than the proportion of white youth in the community. As one might expect, there are also lower proportions of officers who are black, Latino, Asian, and other ethnicities than we see in the youth in the community.

7. Fruit flies.

a) H_0: The ratio of traits in this type of fruit fly is 9:3:3:1, as genetic theory predicts.

H_A: The ratio of traits in this type of fruit fly is not 9:3:3:1.

Counted data condition: The data are counts.
Randomization condition: Assume that these flies are representative of all fruit flies of this type.
Expected cell frequency condition: The expected counts are all greater than 5.

trait	Observed	Expected
YN	59	56.25
YS	20	18.75
EN	11	18.75
ES	10	6.25

Under these conditions, the sampling distribution of the test statistic is χ^2 on $4 - 1 = 3$ degrees of freedom. We will use a chi-square goodness-of-fit test.

$$\chi^2 = \sum_{all\ cells} \frac{(Obs - Exp)^2}{Exp} = \frac{(59-56.25)^2}{56.25} + \frac{(20-18.75)^2}{18.75} + \frac{(11-18.75)^2}{18.75} + \frac{(10-6.25)^2}{6.25} \approx 5.671$$

With $\chi^2 \approx 5.671$, on 3 degrees of freedom, the *P*-value = 0.1288 is high, so we fail to reject the null hypothesis. There is no evidence that the ratio of traits is different than the theoretical ratio predicted by the genetic model. The observed results are consistent with the genetic model.

b) With $\chi^2 \approx 11.342$, on 3 degrees of freedom, the *P*-value = 0.0100 is low, so we reject the null hypothesis. There is strong evidence that the ratio of traits is different than the theoretical ratio predicted by the genetic model. Specifically, there is evidence that the normal wing length occurs less frequently than expected and the short wing length occurs more frequently than expected.

trait	Observed	Expected
YN	118	112.5
YS	40	37.5
EN	22	37.5
ES	20	12.5

c) At first, this seems like a contradiction. We have two samples with exactly the same ratio of traits. The smaller of the two provides no evidence of a difference, yet the larger one provides strong evidence of a difference. This is explained by the sample size. In general, large samples decrease the proportion of variation from the true ratio. Because of the relatively small sample in the first test, we are unwilling to say that there is a difference. There just isn't enough evidence. But the larger sample allows to be more certain about the difference.

9. **Titanic.**

a) $P(\text{crew}) = \frac{885}{2201} \approx 0.402$

b) $P(\text{third and alive}) = \frac{178}{2201} \approx 0.081$

c) $P(\text{alive} \mid \text{first}) = \frac{P(\text{alive and first})}{P(\text{first})} = \frac{202/2201}{325/2201} = \frac{202}{325} \approx 0.622$

d) The overall chance of survival is $\frac{710}{2201} \approx 0.323$, so we would expect about 32.3% of the crew, or about 285.48 members of the crew, to survive.

e) H_0: Survival was independent of status on the ship.

H_A: Survival depended on status on the ship.

f) The table has 2 rows and 4 columns, so there are $(2-1)\times(4-1) = 3$ degrees of freedom.

g) With $\chi^2 \approx 187.8$, on 3 degrees of freedom, the *P*-value is essentially 0, so we reject the null hypothesis. There is strong evidence survival depended on status. First-class passengers were more likely to survive than any other class or crew.

11. **Cranberry juice.**

a) This is an experiment. Volunteers were assigned to treatment groups, each of which drank a different beverage.

b) We are concerned with the proportion of urinary tract infections among three different groups. We will use a chi-square test for homogeneity.

c) H_0: The proportion of urinary tract infection is the same for each group.

H_A: The proportion of urinary tract infection is different among the groups.

d) Counted data condition: The data are counts.
Randomization condition: Although not specifically stated, we will assume that the women were randomly assigned to treatments.
Expected cell frequency condition: The expected counts are all greater than 5.

	Cranberry (Obs / Exp)	Lactobacillus (Obs / Exp)	Control (Obs / Exp)
Infection	8 / 15.333	20 / 15.333	18 / 15.333
No infection	42 / 34.667	30 / 34.667	32 / 34.667

e) The table has 2 rows and 3 columns, so there are $(2-1) \times (3-1) = 2$ degrees of freedom.

f) $\chi^2 = \sum_{all\ cells} \dfrac{(Obs - Exp)^2}{Exp} \approx 7.776$

P-value ≈ 0.020.

g) Since the P-value is low, we reject the null hypothesis. There is strong evidence of difference in the proportion of urinary tract infections for cranberry juice drinkers, lactobacillus drinkers, and women that drink neither of the two beverages.

h) A table of the standardized residuals is below, calculated by using $c = \dfrac{Obs - Exp}{\sqrt{Exp}}$.

	Cranberry	Lactobacillus	Control
Infection	–1.87276	1.191759	0.681005
No infection	1.245505	–0.79259	–0.45291

There is evidence that women who drink cranberry juice are less likely to develop urinary tract infections, and women who drank lactobacillus are more likely to develop urinary tract infections.

13. Montana.

a) We have one group, categorized according to two variables, political party and gender, so we will perform a chi-square test for independence.

b) H_0: Political party is independent of gender in Montana.

H_A: There is an association between political party and gender in Montana.

c) **Counted data condition:** The data are counts.
Randomization condition: Although not specifically stated, we will assume that the poll was conducted randomly.
Expected cell frequency condition: The expected counts are all greater than 5.

	Democrat (Obs/Exp)	Republican (Obs/Exp)	Independent (Obs/Exp)
Male	36 / 43.663	45 / 40.545	24 / 20.792
Female	48 / 40.337	33 / 37.455	16 / 19.208

d) Under these conditions, the sampling distribution of the test statistic is χ^2 on 2 degrees of freedom. We will use a chi-square test for independence.

$$\chi^2 = \sum_{all\ cells} \frac{(Obs - Exp)^2}{Exp} \approx 4.851$$

The *P*-value ≈ 0.0884

e) Since the *P*-value ≈ 0.0884 is fairly high, we fail to reject the null hypothesis. There is little evidence of an association between gender and political party in Montana.

15. Montana revisited.

H_0: Political party is independent of region in Montana.

H_A: There is an association between political party and region in Montana.

Counted data condition: The data are counts.
Randomization condition: Although not specifically stated, we will assume that the poll was conducted randomly.
Expected cell frequency condition: The expected counts are all greater than 5.

	Democrat (Obs/Exp)	Republican (Obs/Exp)	Independent (Obs/Exp)
West	39 / 28.277	17 / 26.257	12 / 13.465
Northeast	15 / 23.703	30 / 22.01	12 / 11.287
Southeast	30 /32.02	31 / 29.733	16 / 15.248

Under these conditions, the sampling distribution of the test statistic is χ^2 on 4 degrees of freedom. We will use a chi-square test for independence.

$$\chi^2 = \sum_{all\ cells} \frac{(Obs - Exp)^2}{Exp} \approx 13.849, \text{ and the } P\text{-value} \approx 0.0078$$

Since the *P*-value ≈ 0.0078 is low, reject the null hypothesis.
There is strong evidence of an association between region and political party in Montana. Residents in the West are more likely to be Democrats than Republicans, and residents in the Northeast are more likely to be Republicans than Democrats.

17. Grades.

a) We have two groups, students of Professor Alpha and students of Professor Beta, and we are concerned with the distribution of one variable, grade. We will perform a chi-square test for homogeneity.

b) H_0: The distribution of grades is the same for the two professors.

H_A: The distribution of grades is different for the two professors.

The expected counts are organized in the table below:

	Prof. Alpha	Prof. Beta
A	6.667	5.333
B	12.778	10.222
C	12.222	9.778
D	6.111	4.889
F	2.222	1.778

Since three cells have expected counts less than 5, the chi-square procedures are not appropriate. Cells would have to be combined in order to proceed. (We will do this in Exercise 19.)

19. Grades again.

a) **Counted data condition:** The data are counts.
Randomization condition: Assume that these students are representative of all students that have ever taken courses from the professors.
Expected cell frequency condition: The expected counts are all greater than 5.

	Prof. Alpha (Obs/Exp)	Prof. Beta (Obs/Exp)
A	3 / 6.667	9 / 5.333
B	11 / 12.778	12 / 10.222
C	14 / 12.222	8 / 9.778
Below C	12 / 8.333	3 / 6.667

b) Under these conditions, the sampling distribution of the test statistic is χ^2 on 3 degrees of freedom, instead of 4 degrees of freedom before the change in the table. We will use a chi-square test for homogeneity.

c) With $\chi^2 = \sum_{all\,cells} \frac{(Obs - Exp)^2}{Exp} \approx 9.306$, the *P*-value ≈ 0.0255.

Since *P*-value = 0.0255 is low, we reject the null hypothesis.
There is evidence that the grade distributions for the two professors are different.
Professor Alpha gives fewer As and more grades below C than Professor Beta.

21. Racial steering.

H_0: There is no association between race and the section of the apartment complex in which people live.

H_A: There is an association between race and the section of the apartment complex in which people live.

Counted data condition: The data are counts.
Randomization condition: Assume that the recently rented apartments are representative of all apartments in the complex.
Expected cell frequency condition: The expected counts are all greater than 5.

	White (Obs / Exp)	Black (Obs / Exp)
Section A	87 / 76.179	8 / 18.821
Section B	83 / 93.821	34 / 23.179

Under these conditions, the sampling distribution of the test statistic is χ^2 on 1 degree of freedom. We will use a chi-square test for independence.

$$\chi^2 = \sum_{all\,cells} \frac{(Obs - Exp)^2}{Exp} \approx \frac{(87 - 76.179)^2}{76.179} + \frac{(8 - 18.821)^2}{18.821} + \frac{(83 - 93.821)^2}{93.821} + \frac{(34 - 23.179)^2}{23.179}$$

$$\approx 1.5371 + 6.2215 + 1.2481 + 5.0517$$

$$\approx 14.058$$

With $\chi^2 \approx 14.058$, on 1 degree of freedom, the P-value ≈ 0.0002.

Since the P-value ≈ 0.0002 is low, we reject the null hypothesis. There is strong evidence of an association between race and the section of the apartment complex in which people live. An examination of the components shows us that whites are much more likely to rent in Section A (component = 6.2215), and blacks are much more likely to rent in Section B (component = 5.0517).

23. Steering revisited.

a) H_0: The proportion of whites who live in Section A is the same as the proportion of blacks who live in Section A. $\left(p_W = p_B \text{ or } p_W - p_B = 0\right)$

H_A: The proportion of whites who live in Section A is different than the proportion of blacks who live in Section A. $\left(p_W \neq p_B \text{ or } p_W - p_B \neq 0\right)$

Independence assumption: Assume that people rent apartments independent of the section.
Independent samples condition: The groups are not associated.
Success/Failure condition: $n\hat{p}$ (white) = 87, $n\hat{q}$ (white) = 83, $n\hat{p}$ (black) = 8, and $n\hat{q}$ (black) = 34. These are not all greater than 10, since the number of black renters in Section A is only 8, but it is close to 10, and the others are large. It should be safe to proceed.

Since the conditions have been satisfied, we will model the sampling distribution of the difference in proportion with a Normal model with mean 0 and standard deviation

estimated by $SE_{pooled}\left(\hat{p}_W - \hat{p}_B\right) = \sqrt{\dfrac{\hat{p}_{pooled}\hat{q}_{pooled}}{n_W} + \dfrac{\hat{p}_{pooled}\hat{q}_{pooled}}{n_B}} = \sqrt{\dfrac{\left(\frac{95}{212}\right)\left(\frac{117}{212}\right)}{170} + \dfrac{\left(\frac{95}{212}\right)\left(\frac{117}{212}\right)}{42}} \approx 0.0856915.$

The observed difference between the proportions is:
0.5117647 – 0.1904762 = 0.3212885.

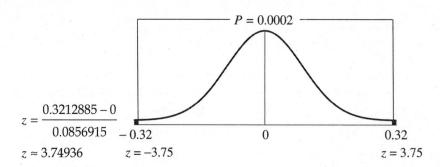

$$z = \frac{0.3212885 - 0}{0.0856915}$$

$$z \approx 3.74936$$

(You have to use a *ridiculous* number of decimal places to get this to come out "right". This is to done to illustrate the point of the question. DO NOT DO THIS! Use technology.)

Since the *P*-value = 0.0002 is low, we reject the null hypothesis. There is strong evidence of a difference in the proportion of whites and blacks living in Section A. The evidence suggests that the proportion of all whites living in Section A is much higher than the proportion of all black residents living in Section A.

The value of z for this test was approximately 3.74936. $z^2 \approx (3.74936)^2 \approx 14.058$, the same as the value for χ^2 in Exercise 21.

b) The resulting *P*-values were both approximately 0.0002. The two tests are equivalent.

25. Race and education.

H_0: Race is independent of education level.

H_A: There is an association between race and education level.

Counted data condition: The data are counts.
Randomization condition: Assume that the sample was taken randomly.
Expected cell frequency condition: The expected counts are all greater than 5.

	Not HS Grad (Obs/Exp)	HS Diploma (Obs/Exp)	College Grad (Obs/Exp)	Adv. Degree (Obs/Exp)
White	810 / 1469	6429 / 6523.8	4725 / 4055.7	1127 / 1042.5
Black	263 / 283.57	1598 / 1259.3	549 / 782.88	117 / 201.24
Hispanic	1031 / 315.43	1269 / 1400.8	412 / 870.86	99 / 223.86
Other	66 / 102	341 / 453	305 / 281.61	197 / 72.389

Under these conditions, the sampling distribution of the test statistic is χ^2 on 9 degrees of freedom. We will use a chi-square test for independence.

With $\chi^2 = \sum_{all\ cells} \frac{(Obs - Exp)^2}{Exp} \approx 2815.968$, on 9 degrees of freedom, the *P*-value is essentially 0.

Since the *P*-value is so low, we reject the null hypothesis. There is strong evidence of an association between race and education level. Hispanics are more likely to not have a high school diploma, and fewer whites fail to graduate from high school than expected.

27. Race and education, part 2.

H_0: Race is independent of education level.

H_A: There is an association between race and education level.

Counted data condition: The data are counts.
Randomization condition: Assume that the sample was taken randomly.
Expected cell frequency condition: The expected counts are all greater than 5.

	HS Diploma (Obs / Exp)	College Grad (Obs / Exp)	Adv. Degree (Obs / Exp)
Black	1598 / 1605.1	549 / 538.01	117 / 120.93
Hispanic	1269 / 1261.9	412 / 422.99	99 / 95.074

Under these conditions, the sampling distribution of the test statistic is χ^2 on 2 degrees of freedom. We will use a chi-square test for independence.

With $\chi^2 = \sum_{all\ cells} \dfrac{(Obs - Exp)^2}{Exp} \approx 0.870$, on 2 degrees of freedom, the P-value ≈ 0.6471.

Since the P-value is high, we fail to reject the null hypothesis. There is no evidence of an association between race and educational performance. Blacks and Hispanics appear to have similar distributions of educational opportunity.

Chapter 27 – Inferences for Regression

1. Ms. President?

a) The equation of the line of best fit for these data points is $\%\hat{Y}es = -5.583 + 0.999(Year)$, where *Year* is measured in years since 1900. According to the linear model, the percentage of people willing to vote for a female candidate for president is increasing by about 1% each year.

b) H_0: There has been no change in the percentage of people willing to vote for a female candidate for president. $(\beta_1 = 0)$

   H_A: There has been a change in the percentage of people willing to vote for a female candidate for president. $(\beta_1 \neq 0)$

c) Assuming the conditions have been met, the sampling distribution of the regression slope can be modeled by a Student's *t*-model with (16 – 2) = 14 degrees of freedom. We will use a regression slope *t*-test.

   The value of *t* = 15.1. The *P*-value ≤ 0.0001 means that the association we see in the data is unlikely to occur by chance. We reject the null hypothesis, and conclude that there is strong evidence that the percentage of people willing to vote for a female presidential candidate is increasing.

d) 94.2% of the variation in the percentage of people willing to vote for a female president is explained by the year.

3. No opinion.

a) H_0: There has been no change in the percentage of people expressing no opinion about their willingness to vote for a female candidate for president. $(\beta_1 = 0)$

   H_A: There has been a change in the percentage of people expressing no opinion about their willingness to vote for a female candidate for president. $(\beta_1 \neq 0)$

b) Assuming the conditions have been met, the sampling distribution of the regression slope can be modeled by a Student's *t*-model with (16 – 2) = 14 degrees of freedom. We will use a regression slope *t*-test.

   The equation of the line of best fit for these data points is:
   $\%No\hat{O}pinion = 7.693 - 0.0427(Year)$, where *Year* is measured in years since 1900.

   The value of *t* = –1.21. The *P*-value of 0.2458 means that the association we see in the data is likely to occur by chance alone. We fail to reject the null hypothesis, and conclude that there is no evidence of a change in the percentage of people expressing no opinion about their willingness to vote for a female presidential candidate.

c) The plot indicates no real trend. The relatively high value in 1945 and the relatively low value around 2000 are influential points. The true slope does not appear to be negative after discounting influential points at the ends. However, this does not change the conclusion of the hypotheses test in part b. We failed to find evidence of a relationship anyway.

5. **Marriage age.**

a) H_0: The difference in age between men and women at first marriage has not been decreasing since 1975. $(\beta_1 = 0)$

   H_A: The difference in age between men and women at first marriage has been decreasing since 1975. $(\beta_1 < 0)$

b) **Straight enough condition:** The scatterplot is not provided, but the residuals plot looks unpatterned. The scatterplot is likely to be straight enough.
   Independence assumption: We are examining a relationship over time, so there is reason to be cautious, but the residuals plot shows no evidence of dependence.
   Does the plot thicken? condition: The residuals plot shows no obvious trends in the spread.
   Nearly Normal condition: The histogram is not particularly unimodal and symmetric, but shows no obvious skewness of outliers.

   Since conditions have been satisfied, the sampling distribution of the regression slope can be modeled by a Student's t-model with $(24 - 2) = 22$ degrees of freedom. We will use a regression slope t-test. The equation of the line of best fit for these data points is:
   $(Men - \hat{Women}) = 49.9 - 0.024(Year)$

   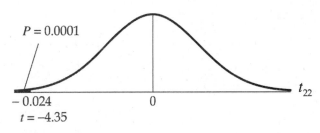

   The value of $t = -4.35$. The P-value of about 0.0001 (remember that regression output usually gives two-sided P-values, so the one-sided P-value = 0.0003/2) means that the association we see in the data is unlikely to occur by chance. We reject the null hypothesis, and conclude that there is strong evidence of a negative linear relationship between difference in age at first marriage and year. The difference in marriage age between men and women appears to be decreasing over time.

d) $b_1 \pm t^*_{n-2} \times SE(b_1) = -0.024 \pm (2.074) \times 0.0055 \approx (-0.035, -0.013)$

   We are 95% confident that the mean difference in age between men and women at first marriage decreases by between 0.013 and 0.035 years in age for each year that passes.

7. **Fuel economy.**

a) H_0: There is no linear relationship between the weight of a car and its mileage. $(\beta_1 = 0)$

   H_A: There is a linear relationship between the weight of a car and its mileage. $(\beta_1 \neq 0)$

b) **Straight enough condition:** The scatterplot is straight enough to try a linear model.
Independence assumption: The residuals plot is scattered.
Does the plot thicken? condition: The residuals plot indicates some possible "thickening" as the predicted values increases, but it's probably not enough to worry about.
Nearly Normal condition: The histogram of residuals is unimodal and symmetric, with one possible outlier. With the large sample size, it is okay to proceed.

Since conditions have been satisfied, the sampling distribution of the regression slope can be modeled by a Student's t-model with (50 – 2) = 48 degrees of freedom. We will use a regression slope t-test. The equation of the line of best fit for these data points is:
$\hat{MPG} = 48.7393 - 8.21362(Weight)$, where *Weight* is measured in thousands of pounds.

The value of $t = -12.2$. The *P*-value of less than 0.0001 means that the association we see in the data is unlikely to occur by chance. We reject the null hypothesis, and conclude that there is strong evidence of a linear relationship between weight of a car and its mileage. Cars that weigh more tend to have lower gas mileage.

9. **Fuel economy, part II.**

a) Since conditions have been satisfied in Exercise 7, the sampling distribution of the regression slope can be modeled by a Student's t-model with (50 – 2) = 48 degrees of freedom. (Use $t_{45}^* = 2.014$ from the table.) We will use a regression slope t-interval, with 95% confidence.

$$b_1 \pm t_{n-2}^* \times SE(b_1) = -8.21362 \pm (2.014) \times 0.6738 \approx (-9.57, -6.86)$$

b) We are 95% confident that the mean mileage of cars decreases by between 6.86 and 9.57 miles per gallon for each additional 1000 pounds of weight.

11. **MPG revisited.**

a) The regression equation predicts that cars that weigh 2500 pounds will have a mean fuel efficiency of $48.7393 - 8.21362(2.5) = 28.20525$ miles per gallon.

$$\hat{y}_v \pm t_{n-2}^* \sqrt{SE^2(b_1) \cdot (x_v - \bar{x})^2 + \frac{s_e^2}{n}}$$

$$= 28.20525 \pm (2.014) \sqrt{0.6738^2 \cdot (2.5 - 2.8878)^2 + \frac{2.413^2}{50}}$$

$$\approx (27.34, 29.07)$$

We are 95% confident that cars weighing 2500 pounds will have mean fuel efficiency between 27.34 and 29.07 miles per gallon.

b) The regression equation predicts that cars that weigh 3450 pounds will have a mean fuel efficiency of $48.7393 - 8.21362(3.45) = 20.402311$ miles per gallon.

$$\hat{y}_v \pm t^*_{n-2}\sqrt{SE^2(b_1)\cdot(x_v - \bar{x})^2 + \frac{s_e^2}{n} + s_e^2}$$

$$= 20.402311 \pm (2.014)\sqrt{0.6738^2 \cdot (3.45 - 2.8878)^2 + \frac{2.413^2}{50} + 2.413^2}$$

$$\approx (15.44,\ 25.37)$$

We are 95% confident that a car weighing 3450 pounds will have fuel efficiency between 15.44 and 25.37 miles per gallon.

13. Cereal.

a) H_0: There is no linear relationship between the number of calories and the sodium content of cereals. $(\beta_1 = 0)$

H_A: There is a linear relationship between the number of calories and the sodium content of cereals. $(\beta_1 \neq 0)$

Since these data were judged acceptable for inference, the sampling distribution of the regression slope can be modeled by a Student's t-model with $(77 - 2) = 75$ degrees of freedom. We will use a regression slope t-test. The equation of the line of best fit for these data points is: $\hat{Sodium} = 21.4143 + 1.29357(Calories)$.

The value of $t = 2.73$. The P-value of 0.0079 means that the association we see in the data is unlikely to occur by chance. We reject the null hypothesis, and conclude that there is strong evidence of a linear relationship between the number of calories and sodium content of cereals. Cereals with higher numbers of calories tend to have higher sodium contents.

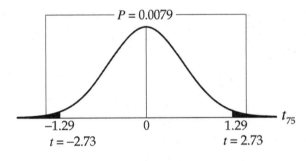

b) Only 9% of the variability in sodium content can be explained by the number of calories. The residual standard deviation is 80.49 mg, which is pretty large when you consider that the range of sodium content is only 320 mg. Although there is strong evidence of a linear association, it is too weak to be of much use. Predictions would tend to be very imprecise.

15. Another bowl.

Straight enough condition: The scatterplot is not straight.
Independence assumption: The residuals plot shows a curved pattern.
Does the plot thicken? condition: The spread of the residuals is not consistent. The residuals plot "thickens" as the predicted values increase.
Nearly Normal condition: The histogram of residuals is skewed to the right, with an outlier.

These data are not appropriate for inference.

17. Acid rain.

a) H₀: There is no linear relationship between BCI and pH. $(\beta_1 = 0)$

Hₐ: There is a linear relationship between BCI and pH. $(\beta_1 \neq 0)$

Assuming the conditions for inference are satisfied, the sampling distribution of the regression slope can be modeled by a Student's *t*-model with (163 – 2) = 161 degrees of freedom. We will use a regression slope *t*-test. The equation of the line of best fit for these data points is: $\hat{BCI} = 2733.37 - 197.694(pH)$.

$$t = \frac{b_1 - \beta_1}{SE(b_1)}$$

$$t = \frac{-197.694 - 0}{25.57}$$

$$t \approx -7.73$$

The value of $t \approx -7.73$. The *P*-value (two-sided!) of essentially 0 means that the association we see in the data is unlikely to occur by chance. We reject the null hypothesis, and conclude that there is strong evidence of a linear relationship between BCI and pH. Streams with higher pH tend to have lower BCI.

19. Ozone.

a) H₀: There is no linear relationship between population and ozone level. $(\beta_1 = 0)$

Hₐ: There is a positive linear relationship between population and ozone level. $(\beta_1 > 0)$

Assuming the conditions for inference are satisfied, the sampling distribution of the regression slope can be modeled by a Student's *t*-model with (16 – 2) = 14 degrees of freedom. We will use a regression slope *t*-test. The equation of the line of best fit for these data points is: $\hat{Ozone} = 18.892 + 6.650(Population)$, where ozone level is measured in parts per million and population is measured in millions.

The value of $t \approx 3.48$. The *P*-value of 0.0018 means that the association we see in the data is unlikely to occur by chance. We reject the null hypothesis, and conclude that there is strong evidence of a positive linear relationship between ozone level and population. Cities with larger populations tend to have higher ozone levels.

$$t = \frac{b_1 - \beta_1}{SE(b_1)}$$

$$t = \frac{6.650 - 0}{1.910}$$

$$t \approx 3.48$$

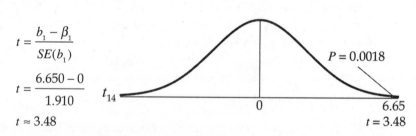

b) $b_1 \pm t_{n-2}^* \times SE(b_1) = 6.65 \pm (1.761) \times 1.910 \approx (3.29, 10.01)$

We are 90% confident that each additional million people will increase mean ozone levels by between 3.29 and 10.01 parts per million.

c) The regression equation predicts that cities with a population of 600,000 people will have ozone levels of 18.892 + 6.650(0.6) = 22.882 parts per million.

$$\hat{y}_v \pm t^*_{n-2}\sqrt{SE^2(b_1)\cdot(x_v - \bar{x})^2 + \frac{s_e^2}{n}}$$

$$= 22.882 \pm (1.761)\sqrt{1.91^2\cdot(0.6-1.7)^2 + \frac{5.454^2}{16}}$$

$$\approx (18.47,\ 27.29)$$

We are 90% confident that the mean ozone level for cities with populations of 600,000 will be between 18.47 and 27.29 parts per million.

21. Start the car!

a) Since the number of degrees of freedom is 33 – 2 = 31, there were 33 batteries tested.

b) **Straight enough condition:** The scatterplot is roughly straight, but very scattered.
Independence assumption: The residuals plot shows no pattern.
Does the plot thicken? condition: The spread of the residuals is consistent.
Nearly Normal condition: The Normal probability plot of residuals is reasonably straight.

c) H_0: There is no linear relationship between cost and power. $(\beta_1 = 0)$

H_A: There is a positive linear relationship between cost and power. $(\beta_1 > 0)$

Since the conditions for inference are satisfied, the sampling distribution of the regression slope can be modeled by a Student's t-model with (33 – 2) = 31 degrees of freedom. We will use a regression slope t-test. The equation of the line of best fit for these data points is: $\widehat{Power} = 384.594 + 4.14649(Cost)$, with power measured in cold cranking amps, and cost measured in dollars.

The value of $t \approx 3.23$. The P-value of 0.0015 means that the association we see in the data is unlikely to occur by chance. We reject the null hypothesis, and conclude that there is strong evidence of a positive linear relationship between cost and power. Batteries that cost more tend to have more power.

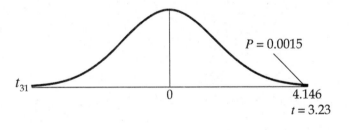

d) Since $R^2 = 25.2\%$, only 25.2% of the variability in power can be explained by cost. The residual standard deviation is 116 amps. That's pretty large, considering the range battery power is only about 400 amps. Although there is strong evidence of a linear association, it is too weak to be of much use. Predictions would tend to be very imprecise.

e) The equation of the line of best fit for these data points is: $\widehat{Power} = 384.594 + 4.14649(Cost)$, with power measured in cold cranking amps, and cost measured in dollars.

f) There are 31 degrees of freedom, so use $t^*_{30} = 1.697$ as a conservative estimate from the table.

$$b_1 \pm t^*_{n-2} \times SE(b_1) = 4.14649 \pm (1.697) \times 1.282 \approx (1.97, 6.32)$$

g) We are 95% confident that the mean power increases by between 1.97 and 6.32 cold cranking amps for each additional dollar in cost.

23. Printers.

H_0: There is no linear relationship between speed and cost of printers. $(\beta_1 = 0)$

H_A: There is a linear relationship between speed and cost of printers. $(\beta_1 \neq 0)$

Straight enough condition: The scatterplot is straight enough to try linear regression, although it looks very scattered, and there doesn't appear to be any association.
Independence assumption: The residuals plot shows no pattern.
Does the plot thicken? condition: The spread of the residuals is consistent
Nearly Normal condition: The Normal probability plot of residuals isn't very straight, and the histogram of the residuals is strongly skewed to the right.

Since we had difficulty with the conditions, inference is not appropriate.
(If you had proceeded with the test, the *P*-value of 0.7756 would have failed to provide evidence of a linear relationship.)

25. Body fat.

a) H_0: There is no linear relationship between waist size and percent body fat. $(\beta_1 = 0)$

H_A: Percent body fat increases with waist size. $(\beta_1 > 0)$

Straight enough condition: The scatterplot is straight enough to try linear regression.
Independence assumption: The residuals plot shows no pattern.
Does the plot thicken? condition: The spread of the residuals is consistent.
Nearly Normal condition: The Normal probability plot of residuals is straight, and the histogram of the residuals is unimodal and symmetric with no outliers.

Since the conditions for inference are inference are satisfied, the sampling distribution of the regression slope can be modeled by a Student's *t*-model with (20 – 2) = 18 degrees of freedom. We will use a regression slope *t*-test.

Dependent variable is: **Body Fat %**
No Selector
R squared = 78.7%   R squared (adjusted) = 77.5%
s = 4.540 with 20 - 2 = 18 degrees of freedom

The equation of the line of best fit for these data points is: $\%\hat{Body}Fat = -62.5573 + 2.22152(Waist)$.

Source	Sum of Squares	df	Mean Square	F-ratio
Regression	1366.79	1	1366.79	66.3
Residual	370.960	18	20.6089	

Variable	Coefficient	s.e. of Coeff	t-ratio	prob
Constant	-62.5573	10.16	-6.16	≤ 0.0001
Waist (in)	2.22152	0.2728	8.14	≤ 0.0001

The value of $t \approx 8.14$. The *P*-value of essentially 0 means that the association we see in the data is unlikely to occur by chance. We reject the null hypothesis, and conclude that there is strong evidence of a linear relationship between waist size and percent body fat. People with larger waists tend to have a higher percentage of body fat.

b) The regression equation predicts that people with 40-inch waists will have $-62.5573 + 2.22152(40) = 20.3035\%$ body fat. The average waist size of the people sampled was approximately 37.05 inches.

$$\hat{y}_v \pm t^*_{n-2}\sqrt{SE^2(b_1)\cdot(x_v - \bar{x})^2 + \frac{s_e^2}{n}}$$

$$= 20.3035 \pm (2.101)\sqrt{0.2728^2 \cdot (40 - 37.05)^2 + \frac{4.54^2}{20}}$$

$$\approx (17.58, 23.03)$$

We are 95% confident that the mean percent body fat for people with 40-inch waists is between 17.58% and 23.03%.

27. Education and mortality.

a) **Straight enough condition:** The scatterplot is straight enough to try linear regression.
Independence assumption: The residuals plot shows no pattern. If these cities are representative of other cities, we can generalize our results.
Does the plot thicken? condition: The spread of the residuals is consistent.
Nearly Normal condition: The histogram of the residuals is unimodal and symmetric with no outliers.

b) H_0: There is no linear relationship between education and mortality. $(\beta_1 = 0)$

H_A: Cities with higher average education level have lower mortality rates. $(\beta_1 < 0)$

Since the conditions for inference are inference are satisfied, the sampling distribution of the regression slope can be modeled by a Student's t-model with $(58 - 2) = 56$ degrees of freedom. We will use a regression slope t-test. The equation of the line of best fit for these data points is: $\widehat{Mortality} = 1493.26 - 49.9202(Education)$.

The value of $t \approx -6.24$. The P-value of essentially 0 means that the association we see in the data is unlikely to occur by chance. We reject the null hypothesis, and conclude that there is strong evidence of a linear relationship between the level of education in a city and its mortality rate. Cities with lower education levels tend to have higher mortality rates.

$$t = \frac{b_1 - \beta_1}{SE(b_1)}$$

$$t = \frac{-49.9202 - 0}{8.000}$$

$$t \approx -6.24$$

c) For 95% confidence, $t^*_{56} \approx 2.00327$.

$$b_1 \pm t^*_{n-2} \times SE(b_1) = -49.9202 \pm (2.003) \times 8.000 \approx (-65.95, -33.89)$$

d) We are 95% confident that the mean number of deaths per 100,000 people decreases by between 33.89 and 65.95 deaths for an increase of one year in average education level.

e) The regression equation predicts that cities with an adult population with an average of 12 years of school will have a mortality rate of $1493.26 - 49.9202(12) = 894.2176$ deaths per 100,000. The average education level was 11.0328 years.

$$\hat{y}_v \pm t^*_{n-2} \sqrt{SE^2(b_1) \cdot (x_v - \bar{x})^2 + \frac{s_e^2}{n}}$$

$$= 894.2176 \pm (2.003) \sqrt{8.00^2 \cdot (12 - 11.0328)^2 + \frac{47.92^2}{58}}$$

$$\approx (874.239, 914.196)$$

We are 95% confident that the mean mortality rate for cities with an average of 12 years of schooling is between 874.239 and 914.196 deaths per 100,000 residents.

Review of Part VII – Inference When Variables Are Related

1. Genetics.

H_0: The proportions of traits are as specified by the ratio 1:3:3:9.
H_A: The proportions of traits are not as specified.

Counted data condition: The data are counts.
Randomization condition: Assume that these students are representative of all people.
Expected cell frequency condition: The expected counts (shown in the table) are all greater than 5.

Under these conditions, the sampling distribution of the test statistic is χ^2 on $4 - 1 = 3$ degrees of freedom. We will use a chi-square goodness-of-fit test.

Trait	Observed	Expected	Residual = $(Obs - Exp)$	$(Obs - Exp)^2$	Component = $\dfrac{(Obs - Exp)^2}{Exp}$
Attached, noncurling	10	7.625	2.375	5.6406	0.73975
Attached, curling	22	22.875	-0.875	0.7656	0.03347
Free, noncurling	31	22.875	8.125	66.0156	2.8859
Free, curling	59	68.625	-9.625	92.6406	1.35

$$\sum \approx 5.01$$

$\chi^2 = 5.01$. Since the *P*-value = 0.1711 is high, we fail to reject the null hypothesis.

There is no evidence that the proportions of traits are anything other than 1:3:3:9.

3. Hard water.

a) H_0: There is no linear relationship between calcium concentration in water and mortality rates for males. $(\beta_1 = 0)$
H_A: There is a linear relationship between calcium concentration in water and mortality rates for males. $(\beta_1 \neq 0)$

b) Assuming the conditions have been satisfied, the sampling distribution of the regression slope can be modeled by a Student's t-model with $(61 - 2) = 59$ degrees of freedom. We will use a regression slope t-test. The equation of the line of best fit for these data points is: $\widehat{Mortality} = 1676 - 3.23(Calcium)$, where mortality is measured in deaths per 100,000, and calcium concentration is measured in parts per million.

$$t = \frac{b_1 - \beta_1}{SE(b_1)}$$

$$t = \frac{-3.23 - 0}{0.48}$$

$$t \approx -6.73$$

The value of $t = -6.73$. The P-value of less than 0.0001 means that the association we see in the data is unlikely to occur by chance. We reject the null hypothesis, and conclude that there is strong evidence of a linear relationship between calcium concentration and mortality. Towns with higher calcium concentrations tend to have lower mortality rates.

c) For 95% confidence, use $t_{59}^* \approx 2.001$, or estimate from the table $t_{50}^* \approx 2.009$.

$$b_1 \pm t_{n-2}^* \times SE(b_1) = -3.23 \pm (2.001) \times 0.48 \approx (-4.19, -2.27)$$

d) We are 95% confident that the average mortality rate decreases by between 2.27 and 4.19 deaths per 100,000 for each additional part per million of calcium in drinking water.

5. Resume fraud.

In order to estimate the true percentage of people have misrepresented their backgrounds to within ± 5%, the company would have to perform about 406 random checks.

$$ME = z^* \sqrt{\frac{\hat{p}\hat{q}}{n}}$$

$$0.05 = 2.326 \sqrt{\frac{(0.25)(0.75)}{n}}$$

$$n = \frac{(2.326)^2(0.25)(0.75)}{(0.05)^2}$$

$$n \approx 406 \text{ random checks}$$

7. Back to Montana.

H_0: Political party is independent of income level in Montana.

H_A: There is an association between political party and income level in Montana.

Counted data condition: The data are counts.
Randomization condition: Although not specifically stated, we will assume that the poll was conducted randomly.
Expected cell frequency condition: The expected counts are all greater than 5.

	Democrat (Obs/Exp)	Republican (Obs/Exp)	Independent (Obs/Exp)
Low	30 / 24.119	16 / 22.396	12 / 14.485
Middle	28 / 30.772	24 / 28.574	22 / 14.653
High	26 / 29.109	38 / 27.03	6 / 13.861

Under these conditions, the sampling distribution of the test statistic is χ^2 on 4 degrees of freedom. We will use a chi-square test for independence.

$$\chi^2 = \sum_{\text{all cells}} \frac{(Obs - Exp)^2}{Exp} \approx 17.19$$

The *P*-value ≈ 0.0018

Since the *P*-value ≈ 0.0018 is low, we reject the null hypothesis. There is strong evidence of an association between income level and political party in Montana. An examination of the components shows that Democrats are more likely to have low incomes, Independents are more likely to have middle incomes, and Republicans are more likely to have high incomes.

9. **Lefties and music.**

   H_0: The proportion of right-handed people who can match the tone is the same as the proportion of left-handed people who can match the tone. $(p_L = p_R \text{ or } p_L - p_R = 0)$

   H_A : The proportion of right-handed people who can match the tone is different from the proportion of left-handed people who can match the tone. $(p_L \neq p_R \text{ or } p_L - p_R \neq 0)$

   Random condition: Assume that the people tested are representative of all people.
   10% condition: 76 and 53 are both less than 10% of all people.
   Independent samples condition: The groups are not associated.
   Success/Failure condition: $n\hat{p}$ (right) = 38, $n\hat{q}$ (right) = 38, $n\hat{p}$ (left) = 33, and $n\hat{q}$ (left) = 20 are all greater than 10, so the samples are both large enough.

   Since the conditions have been satisfied, we will model the sampling distribution of the difference in proportion with a Normal model with mean 0 and standard deviation

   estimated by $SE_{\text{pooled}}(\hat{p}_L - \hat{p}_R) = \sqrt{\dfrac{\hat{p}_{\text{pooled}}\hat{q}_{\text{pooled}}}{n_L} + \dfrac{\hat{p}_{\text{pooled}}\hat{q}_{\text{pooled}}}{n_R}} = \sqrt{\dfrac{\left(\frac{71}{129}\right)\left(\frac{58}{129}\right)}{53} + \dfrac{\left(\frac{71}{129}\right)\left(\frac{58}{129}\right)}{76}} \approx 0.089.$

   The observed difference between the proportions is:
   0.6226 – 0.5 = 0.1226.

   Since the *P*-value = 0.1683 is high, we fail to reject the null hypothesis. There is no evidence that the proportion of people able to match the tone differs between right-handed and left-handed people.

   $z = \dfrac{0.1226 - 0}{0.089}$

   $z \approx 1.38$

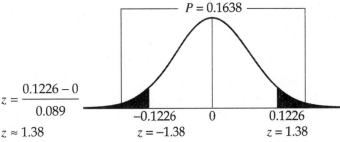

11. **Polling.**

    a) H_0: The mean difference in the number of predicted Democrats and the number of actual Democrats is zero. $(\mu_d = 0)$

    H_A: The mean difference in the number of predicted Democrats and the number of actual Democrats is different than zero. $(\mu_d \neq 0)$

Paired data assumption: The data are paired by year.
Randomization condition: Assume these predictions are representative of other predictions.
10% condition: We are testing the predictions, not the years.
Nearly Normal condition: The histogram of differences between the predicted number of Democrats and the actual number of Democrats is roughly unimodal and symmetric. The year 1958 is an outlier, and was removed.

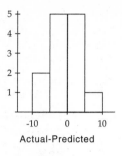

Since the conditions are satisfied, the sampling distribution of the difference can be modeled with a Student's t-model with $13 - 1 = 12$ degrees of freedom, $t_{12}\left(0, \dfrac{3.57878}{\sqrt{13}}\right)$.

We will use a paired t-test, with $\bar{d} = -0.846$.

Since the P-value = 0.4106 is high, we fail to reject the null hypothesis. There is no evidence that the mean difference between the actual and predicted number of Democrats was anything other than 0.

$$t = \frac{\bar{d} - 0}{\frac{s_d}{\sqrt{n}}}$$

$$t = \frac{-0.846 - 0}{\frac{3.57878}{\sqrt{13}}}$$

$$t \approx -0.85$$

b) H_0: There is no linear relationship between Gallup's predictions and the actual number of Democrats. $(\beta_1 = 0)$

H_A: There is a linear relationship between Gallup's predictions and the actual number of Democrats. $(\beta_1 \neq 0)$

Straight enough condition: The scatterplot is straight enough to try linear regression.
Independence assumption: The residuals plot shows no pattern.
Does the plot thicken? condition: The spread of the residuals is consistent.
Nearly Normal condition: After an outlier in 1958 is removed, the Normal probability plot of residuals still isn't very straight. However, the histogram of residuals is roughly unimodal and symmetric. With a sample size of 13, it is probably okay to proceed.

Since the conditions for inference are satisfied, the sampling distribution of the regression slope can be modeled by a Student's t-model with $(13 - 2) = 11$ degrees of freedom. We will use a regression slope t-test.

Dependent variable is: **Actual**
No Selector
R squared = 98.7%   R squared (adjusted) = 98.6%
s = 3.628 with 13 - 2 = 11 degrees of freedom

Source	Sum of Squares	df	Mean Square	F-ratio
Regression	10874.4	1	10874.4	826
Residual	144.805	11	13.1641	

Variable	Coefficient	s.e. of Coeff	t-ratio	prob
Constant	6.00180	8.395	0.715	0.4895
Predicted	0.972206	0.0338	28.7	≤ 0.0001

The equation of the line of best fit for these data points is: $\hat{Actual} = 6.00180 + 0.972206(Predicted)$.

The value of $t \approx 28.7$. The P-value of essentially 0 means that the association we see in the data is unlikely to occur by chance. We reject the null hypothesis, and conclude that there is strong evidence of a linear relationship between the number of Democrats predicted by Gallup and the number of Democrats actually in the House of Representatives. Years in which the predicted number was high tend to have high actual numbers also. The high value of $R^2 = 98.7\%$ indicates a very strong model. Gallup polls seem very accurate.

13. Twins, again.

H_0: The distributions of pregnancy durations are the same for the three years.

H_A: The distributions of pregnancy durations are different for the three years.

Counted data condition: The data are counts.
Independence assumption: Assume that the durations of the pregnancies are mutually indpendent.
Expected cell frequency condition: The expected counts are all greater than 5.

	1990 (Obs/Exp)	1995 (Obs/Exp)	2000 (Obs/Exp)
Preterm (induced or Cesarean)	11 /12.676	13 / 13.173	19 / 17.150
Preterm (without procedures)	13 / 13.266	14 / 13.786	18 / 17.948
Term or postterm	27 / 25.058	26 / 26.04	32 / 33.902

Under these conditions, the sampling distribution of the test statistic is χ^2 on 4 degrees of freedom. We will use a chi-square test for homogeneity.

$$\chi^2 = \sum_{all\ cells} \frac{(Obs - Exp)^2}{Exp} \approx 0.69,$$ and the P-value ≈ 0.9526.

Since the P-value ≈ 0.9526 is high, we fail to reject the null hypothesis. There is no evidence that the distributions of the durations of pregnancies are different for the three years. It does not appear that they way the hospital deals with twin pregnancies has changed.

15. LA rainfall.

a) **Independence assumption:** Annual rainfall is independent from year to year.
Nearly Normal condition: The histogram of the rainfall totals is skewed to the right, but the sample is fairly large, so it is safe to proceed.

The mean annual rainfall is 14.5165 inches, with a standard deviation 7.82044 inches. Since the conditions have been satisfied, construct a one-sample *t*-interval, with 22 – 1 = 21 degrees of freedom, at 90% confidence.

$$\bar{y} \pm t^*_{n-1}\left(\frac{s}{\sqrt{n}}\right) = 14.5164 \pm t^*_{21}\left(\frac{7.82044}{\sqrt{22}}\right) \approx (11.65, 17.39)$$

We are 90% confident that the mean annual rainfall in LA is between 11.65 and 17.39 inches.

b) Start by making an estimate, either using $z^* = 1.645$ or $t^*_{21} = 1.721$ from above. Either way, your estimate is around 40 people. Make a better estimate using $t^*_{40} = 1.684$. You would need about 44 years' data to estimate the annual rainfall in LA to within 2 inches.

$$ME = t^*_{40}\left(\frac{s}{\sqrt{n}}\right)$$

$$2 = 1.684\left(\frac{7.82044}{\sqrt{n}}\right)$$

$$n = \frac{(2.064)^2(7.82044)^2}{(2)^2}$$

$$n \approx 44 \text{ years}$$

c) H_0: There is no linear relationship between year and annual LA rainfall. $(\beta_1 = 0)$
H_A: There is a linear relationship between year and annual LA rainfall. $(\beta_1 \neq 0)$

Straight enough condition: The scatterplot is straight enough to try linear regression, although there is no apparent pattern.
Independence assumption: The residuals plot shows no pattern.
Does the plot thicken? condition: The spread of the residuals is consistent.
Nearly Normal condition: The Normal probability plot of residuals is not straight, and the histogram of the residuals is skewed to the right, but a sample of 22 years is large enough to proceed.

Since the conditions for inference are satisfied, the sampling distribution of the regression slope can be modeled by a Student's *t*-model with (22 – 2) = 20 degrees of freedom. We will use a regression slope *t*-test.

Dependent variable is: **Rain** (in.)
No Selector
R squared = 0.1%   R squared (adjusted) = -4.9%
s = 8.011 with 22 - 2 = 20 degrees of freedom

Source	Sum of Squares	df	Mean Square	F-ratio
Regression	0.979449	1	0.979449	0.015
Residual	1283.37	20	64.1684	

Variable	Coefficient	s.e. of Coeff	t-ratio	prob
Constant	-			
Year	C			

The equation of the line of best fit for these data points is: $\hat{Rain} = -51.6838 + 0.033258(Year)$.

The ... we ... by ...

value of $t \approx 0.124$. The *P*-value of 0.92029 means that the association see in the data is quite likely to occur chance. We fail to reject the null hypothesis, and conclude that there is no evidence of a linear relationship between the annual rainfall in LA and the year.

17. Eye and hair color.

a) This is an attempt at linear regression. Regression inference is meaningless here, since eye and hair color are categorical variables.

b) This is an analysis based upon a chi-square test for independence.

H_0: Eye color and hair color are independent.

H_A: There is an association between eye color and hair color.

Since we have two categorical variables, this analysis seems appropriate. However, if you check the expected counts, you will find that 4 of them are less than 5. We would have to combine several cells in order to perform the analysis. (Always check the conditions!)

Since the value of chi-square is so high, it is likely that we would find an association between eye and hair color, even after the cells were combined. There are many cells of interest, but some of the most striking differences that would not be affected by cell combination involve people with fair hair. Blonds are likely to have blue eyes, and not likely to have brown eyes. Those with red hair are not likely to have brown eyes. Additionally, those with black hair are much more likely to have brown eyes than blue.

19. Pregnancy.

a) H_0: The proportion of live births is the same for women under the age of 38 as it is for women over the age of 38. $\left(p_{<38} = p_{\geq 38} \text{ or } p_{<38} - p_{\geq 38} = 0\right)$

H_A: The proportion of live births is different for women under the age of 38 than for women over the age of 38. $\left(p_{<38} \neq p_{\geq 38} \text{ or } p_{<38} - p_{\geq 38} \neq 0\right)$

Random condition: Assume that the women studied are representative of all women.
10% condition: 157 and 89 are both less than 10% of all women.
Independent samples condition: The groups are not associated.
Success/Failure condition: $n\hat{p}$ (under 38) = 42, $n\hat{q}$ (under 38) = 115, $n\hat{p}$ (38 and over) = 7, and $n\hat{q}$ (38 and over) = 82 are not all greater than 10, since the observed number of live births is only 7. However, if we check the pooled value, $n\hat{p}_{pooled}$ (38 and over) = (89)(0.191) = 17. All of the samples are large enough.

Since the conditions have been satisfied, we will model the sampling distribution of the difference in proportion with a Normal model with mean 0 and standard deviation

estimated by $SE_{pooled}\left(\hat{p}_{<38} - \hat{p}_{\geq38}\right) = \sqrt{\dfrac{\hat{p}_{pooled}\hat{q}_{pooled}}{n_{<38}} + \dfrac{\hat{p}_{pooled}\hat{q}_{pooled}}{n_{\geq38}}} = \sqrt{\dfrac{\left(\frac{49}{246}\right)\left(\frac{197}{246}\right)}{157} + \dfrac{\left(\frac{49}{246}\right)\left(\frac{197}{246}\right)}{89}} \approx 0.0530.$

The observed difference between the proportions is:
$0.2675 - 0.0787 = 0.1888.$

Since the *P*-value = 0.0004 is low, we reject the null hypothesis. There is strong evidence to suggest a difference in the proportion of live births for women under 38 and women 38 and over at this clinic. In fact, the evidence suggests that women under 38 have a higher proportion of live births.

$z = \dfrac{0.1888 - 0}{0.0530}$

$z \approx 3.56$

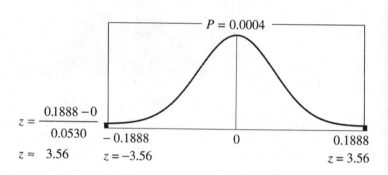

b) H_0: Age and birth rate are independent.

H_A: There is an association between age and birth rate

Counted data condition: The data are counts.
Randomization condition: Assume that these women are representative of all women.
Expected cell frequency condition: The expected counts are all greater than 5.

	Live birth (Obs/Exp)	No live birth (Obs/Exp)
Under 38	42 / 31.272	115 / 125.73
38 and over	7 / 17.728	82 / 71.27

Under these conditions, the sampling distribution of the test statistic is χ^2 on 1 degree of freedom. We will use a chi-square test for independence.

$\chi^2 = \sum_{all\ cells} \dfrac{(Obs - Exp)^2}{Exp} \approx 12.70$, and the *P*-value ≈ 0.0004.

Since the *P*-value ≈ 0.0004 is low, we reject the null hypothesis. There is strong evidence of an association between age and birth rate. Younger mothers tend to have higher birth rates.

c) A two-proportion *z*-test and a chi-square test for independence with 1 degree of freedom are equivalent. $z^2 = (3.563944)^2 = 12.70 = \chi^2$. The *P*-values are both the same.

21. Old Faithful.

a) There is a moderate, linear, positive association between duration of the previous eruption and interval between eruptions for Old Faithful. Relatively long eruptions appear to be associated with relatively long intervals until the next eruption.

b) H₀: There is no linear relationship between duration of the eruption and interval until the next eruption. $(\beta_1 = 0)$
H_A: There is a linear relationship between duration of the eruption and interval until the next eruption. $(\beta_1 \neq 0)$

c) **Straight enough condition:** The scatterplot is straight enough to try linear regression.
Independence assumption: The residuals plot shows no pattern.
Does the plot thicken? condition: The spread of the residuals is consistent.
Nearly Normal condition: The histogram of residuals is unimodal and symmetric.

Since the conditions for inference are satisfied, the sampling distribution of the regression slope can be modeled by a Student's t-model with $(222 - 2) = 220$ degrees of freedom. We will use a regression slope t-test. The equation of the line of best fit for these data points is:
$\widehat{Interval} = 33.9668 + 10.3582(Duration)$.

d) The value of $t \approx 27.1$. The P-value of essentially 0 means that the association we see in the data is unlikely to occur by chance. We reject the null hypothesis, and conclude that there is strong evidence of a linear relationship between duration and interval. Relatively long eruptions tend to be associated with relatively long intervals until the next eruption.

e) The regression equation predicts that an eruption with duration of 2 minutes will have an interval until the next eruption of $33.9668 + 10.3582(2) = 54.6832$ minutes. $(t^*_{220} \approx 1.9708)$

$$\hat{y}_v \pm t^*_{n-2}\sqrt{SE^2(b_1)\cdot(x_v - \bar{x})^2 + \frac{s_e^2}{n}}$$

$$= 54.6832 \pm (1.9708)\sqrt{0.3822^2 \cdot (2 - 3.57613)^2 + \frac{6.159^2}{222}}$$

$$\approx (53.24, 56.12)$$

We are 95% confident that, after a 2-minute eruption, the mean length of time until the next eruption will be between 53.24 and 56.12 minutes.

f) The regression equation predicts that an eruption with duration of 4 minutes will have an interval until the next eruption of $33.9668 + 10.3582(4) = 75.3996$ minutes. $(t^*_{220} \approx 1.9708)$

$$\hat{y}_v \pm t^*_{n-2}\sqrt{SE^2(b_1)\cdot(x_v - \bar{x})^2 + \frac{s_e^2}{n} + s_e^2}$$

$$= 75.3996 \pm (1.9708)\sqrt{0.3822^2 \cdot (4 - 3.57613)^2 + \frac{6.159^2}{222} + 6.159^2}$$

$$\approx (63.23, 87.57)$$

We are 95%confident that the length of time until the next eruption will be between 63.23 and 87.57 minutes, following a 4-minute eruption.

23. Learning math.

a) H₀: The mean score of Accelerated Math students is the same as the mean score of traditional students. $\left(\mu_A = \mu_T \text{ or } \mu_A - \mu_T = 0\right)$

H_A: The mean score of Accelerated Math students is different from the mean score of traditional students. $\left(\mu_A \neq \mu_T \text{ or } \mu_A - \mu_T \neq 0\right)$

Independent groups assumption: Scores of students from different classes should be independent.
Randomization condition: Although not specifically stated, classes in this experiment were probably randomly assigned to learn either Accelerated Math or traditional curricula.
10% condition: 231 and 245 are less than 10% of all students.
Nearly Normal condition: We don't have the actual data, so we can't check the distribution of the sample. However, the samples are large. The Central Limit Theorem allows us to proceed.

Since the conditions are satisfied, it is appropriate to model the sampling distribution of the difference in means with a Student's *t*-model, with 459.24 degrees of freedom (from the approximation formula).

We will perform a two-sample *t*-test. The sampling distribution model has mean 0, with standard error: $SE(\bar{y}_A - \bar{y}_T) = \sqrt{\dfrac{84.29^2}{231} + \dfrac{74.68^2}{245}} \approx 7.3158$.

The observed difference between the mean scores is 560.01 − 549.65 = 10.36

Since the *P*-value = 0.1574, we fail to reject the null hypothesis. There is no evidence that the Accelerated Math students have a different mean score on the pretest than the traditional students.

$$t = \frac{(\bar{y}_A - \bar{y}_T) - (0)}{SE(\bar{y}_A - \bar{y}_T)}$$

$$t \approx \frac{10.36}{7.3158}$$

$$t \approx 1.42$$

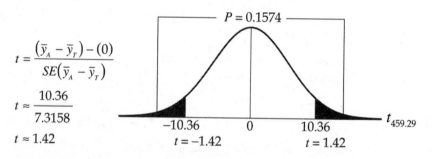

b) H₀: Accelerated Math students do not show significant improvement in test scores. The mean individual gain for Accelerated Math is zero. $\left(\mu_d = 0\right)$

H_A: Accelerated Math students show significant improvement in test scores. The mean individual gain for Accelerated Math is greater than zero. $\left(\mu_d > 0\right)$

Paired data assumption: The data are paired by student.
Randomization condition: Although not specifically stated, classes in this experiment were probably randomly assigned to learn either Accelerated Math or traditional curricula.
10% condition: We are testing the Accelerated Math program, not the students.
Nearly Normal condition: We don't have the actual data, so we cannot look at a graphical display, but since the sample is large, it is safe to proceed.

The Accelerated Math students had a mean individual gain of $\bar{d} = 77.53$ points and a standard deviation of 78.01 points. Since the conditions for inference are satisfied, we can model the sampling distribution of the mean individual gain with a Student's t model, with $231 - 1 = 230$ degrees of freedom, $t_{230}\left(0, \frac{78.01}{\sqrt{231}}\right)$. We will perform a paired t-test.

$$t = \frac{\bar{d} - 0}{\frac{s_d}{\sqrt{n}}}$$

Since the P-value is essentially 0, we reject the null hypothesis. There is strong evidence that the mean individual gain is greater than zero. The Accelerated Math students showed significant improvement.

$$t = \frac{77.53 - 0}{\frac{78.01}{\sqrt{231}}}$$

$$t \approx 15.11$$

c) H_0: Students taught using traditional methods do not show significant improvement in test scores. The mean individual gain for traditional methods is zero. $(\mu_d = 0)$

H_A: Students taught using traditional methods show significant improvement in test scores. The mean individual gain for traditional methods is greater than zero. $(\mu_d > 0)$

Paired data assumption: The data are paired by student.
Randomization condition: Although not specifically stated, classes in this experiment were probably randomly assigned to learn either Accelerated Math or traditional curricula.
10% condition: We are testing the program, not the students.
Nearly Normal condition: We don't have the actual data, so we cannot look at a graphical display, but since the sample is large, it is safe to proceed.

The students taught using traditional methods had a mean individual gain of $\bar{d} = 39.11$ points and a standard deviation of 66.25 points. Since the conditions for inference are satisfied, we can model the sampling distribution of the mean individual gain with a Student's t model, with $245 - 1 = 244$ degrees of freedom, $t_{244}\left(0, \frac{66.25}{\sqrt{245}}\right)$.

We will perform a paired t-test.

Since the P-value is essentially 0, we reject the null hypothesis. There is strong evidence that the mean individual gain is greater than zero. The students taught using traditional methods showed significant improvement.

$$t = \frac{\bar{d} - 0}{\frac{s_d}{\sqrt{n}}}$$

$$t = \frac{39.11 - 0}{\frac{66.25}{\sqrt{245}}}$$

$$t \approx 9.24$$

d) H_0: The mean individual gain of Accelerated Math students is the same as the mean individual gain of traditional students. $(\mu_{dA} = \mu_{dT} \text{ or } \mu_{dA} - \mu_{dT} = 0)$

H_A: The mean individual gain of Accelerated Math students is greater than the mean individual gain of traditional students. $(\mu_{dA} > \mu_{dT} \text{ or } \mu_{dA} - \mu_{dT} > 0)$

Independent groups assumption: Individual gains of students from different classes should be independent.

Randomization condition: Although not specifically stated, classes in this experiment were probably randomly assigned to learn either Accelerated Math or traditional curricula.

10% condition: 231 and 245 are less than 10% of all students.

Nearly Normal condition: We don't have the actual data, so we can't check the distribution of the sample. However, the samples are large. The Central Limit Theorem allows us to proceed.

Since the conditions are satisfied, it is appropriate to model the sampling distribution of the difference in means with a Student's t-model, with 452.10 degrees of freedom (from the approximation formula).

We will perform a two-sample t-test. The sampling distribution model has mean 0, with

standard error: $SE(\bar{d}_A - \bar{d}_T) = \sqrt{\dfrac{78.01^2}{231} + \dfrac{66.25^2}{245}} \approx 6.6527$.

The observed difference between the mean scores is 77.53 – 39.11 = 38.42

$t = \dfrac{(\bar{d}_A - \bar{d}_T) - 0}{SE(\bar{d}_A - \bar{d}_T)}$

Since the P-value is less than 0.0001, we reject the null hypothesis. There is strong evidence that the Accelerated Math students have an individual gain that is significantly higher than the individual gain of the students taught using traditional methods.

$t = \dfrac{38.42 - 0}{6.6527}$

$t \approx 5.78$

25. Dairy sales.

a) Since the CEO is interested in the association between cottage cheese sales and ice cream sales, the regression analysis is appropriate.

b) There is a moderate, linear, positive association between cottage cheese and ice cream sales. For each additional million pounds of cottage cheese sold, an average of 1.19 million pounds of ice cream are sold.

c) The regression will not help here. A paired t-test will tell us whether there is an average difference in sales.

d) There is evidence that the company sells more cottage cheese than ice cream, on average.

e) In part a, we are assuming that the relationship is linear, that errors are independent with constant variation, and that the distribution of errors is Normal.

In part c, we are assuming that the observations are independent and that the distribution of the differences is Normal. This may not be a valid assumption, since the histogram of differences looks bimodal.

f) The equation of the regression line is $\widehat{IceCream} = -26.5306 + 1.19334(CottageCheese)$. In a month in which 82 million pounds of ice cream are sold we expect to sell:

$\widehat{IceCream} = -26.5306 + 1.19334(82) = 71.32$ million pounds of ice cream.

g) Assuming the conditions for inference are satisfied, the sampling distribution of the regression slope can be modeled by a Student's *t*-model with (12 – 2) = 10 degrees of freedom. We will use a regression slope *t*-interval, with 95% confidence.

$$b_1 \pm t^*_{n-2} \times SE(b_1) = 1.19334 \pm (2.228) \times 0.4936 \approx (0.09, 2.29)$$

h) We are 95% confident that the mean number of pounds of ice cream sold increases by between 0.09 and 2.29 pounds for each additional pound of cottage cheese sold.

27. Weight loss.

Randomization Condition: The respondents were randomly selected from among the clients of the weight loss clinic.
10% Condition: 20 people are less than 10% of all clients.
Nearly Normal Condition: The histogram of the number of pounds lost for each respondent is unimodal and symmetric, with no outliers.

The clients in the sample had a mean weight loss of 9.15 pounds, with a standard deviation of 1.94733 pounds. Since the conditions have been satisfied, construct a one-sample *t*-interval, with 20 – 1 = 19 degrees of freedom, at 95% confidence.

$$\bar{y} \pm t^*_{n-1}\left(\frac{s}{\sqrt{n}}\right) = 9.15 \pm t^*_{19}\left(\frac{1.94733}{\sqrt{20}}\right) \approx (8.24, 10.06)$$

We are 95% confident that the mean weight loss experienced by clients of this clinic is between 8.24 and 10.06 pounds. Since 10 pounds is contained within the interval, the claim that the program will allow clients to lose 10 pounds in a month is plausible. Answers may vary, depending on the chosen level of confidence.

29. Diet.

H_0: Cracker type and bloating are independent.

H_A: There is an association between cracker type and bloating.

Counted data condition: The data are counts.
Randomization condition: Assume that these women are representative of all women.
Expected cell frequency condition: The expected counts are all (almost!) greater than 5.

	Bloat	
	Little/None (*Obs / Exp*)	Moderate/Severe (*Obs / Exp*)
Bran	11 / 7.6471	2 / 5.3529
Gum Fiber	4 / 7.6471	9 / 5.3529
Combination	7 / 7.6471	6 / 5.3529
Control	8 / 7.0588	4 / 4.9412

Under these conditions, the sampling distribution of the test statistic is χ^2 on 3 degrees of freedom. We will use a chi-square test for independence.

$$\chi^2 = \sum_{all\ cells} \frac{(Obs - Exp)^2}{Exp} \approx 8.23, \text{ and the } P\text{-value} \approx 0.0414.$$

Since the *P*-value is low, we reject the null hypothesis. There is evidence of an association between cracker type and bloating. The gum fiber crackers had a higher rate of moderate/severe bloating than expected. The company should head back to research and development and address the problem before attempting to market the crackers.